THE

REDUCER'S MANUAL,

AND

Gold & Silver Worker's Guide,

Being a complete, practical Hand-Book on the saving and reduction of every class of

PHOTOGRAPHIC WASTES,

AND

GOLD AND SILVER RESIDUES.

Comprising all the Wet and Dry Processes of Reduction at present known, with many important original Designs, Formulas and Additions.

BY

VICTOR G. BLOEDE,

Late Assistant Chemist at the Cooper Institute.

NEW YORK:
JOSEPH H. LADD, PUBLISHER, No. 91 WHITE STREET.
LONDON: TRUBNER & CO.

DEDICATED

TO

PROF. CHARLES S. STONE,

OF NEW YORK,

AS A SLIGHT TOKEN OF THE ESTEEM OF

THE AUTHOR.

CONTENTS.

CHAPTER I.

CHAPTER II.

CHAPTER III.

CHAPTER IV.

CHATER VII.

CHAPTER VIII.

PART II.

CHAPTER I.

CHAPTER II.

CHAPTER III.

INTRODUCTION.

The great rapidity with which photography, as a business, has grown within the past few years, has been naturally followed by the consumption of immense quantities of the precious metals and their compounds. Probably few photographers know that but a small portion of the precious salts is actually used in the production of the picture, and that more than three-quarters of these valuable compounds still find their way to the waste-pipe. Yet such is the fact. An eminent French chemist, in a "Treatise on Silver and its Compounds," has even taken pains to show us that *nine-tenths* or ninety per cent. of the silver used is actually *lost*, while only one-tenth or ten per cent. is utilized in the production of the picture. Though these figures may be a little overdrawn, still it is a fact beyond question that thousands of dollars are annually lost, merely from want of proper knowledge and attention to this important subject. To assist the practical photographer in accomplishing this utilization of the precious compounds in the most perfect and economical manner, has been our chief object in penning this work. For such as prefer to have their reductions performed by outside parties, the seventh chapter of the MANUAL was written, which gives them a sure defense against fraud or imposition.

I have carefully avoided all technical expressions and theoretical processes, but have confined myself to strictly practical subjects, and feel confident that the work will richly repay the practical operator for the trouble of perusal, and fill a vacancy upon his shelves which he has long and painfully felt.

THE AUTHOR

BROOKLYN, N. Y., May, 1867.

THE

REDUCER'S MANUAL.

CHAPTER I.

HISTORY AND CHEMICAL PROPERTIES OF THE PRECIOUS METALS.*

THE word *element* is understood at the present time as indicating every substance which chemists have not succeeded in decomposing or separating into different component parts. All the metals at present known, both the common and the precious ones, have this characteristic and are consequently termed elements. Gold, silver and platinum are called the noble metals, from the fact that they do not associate with the common elements, but prefer to shine alone in their inborn glory. One of the commonest and most widely diffused of the elements is a constituent of the air we breathe, and this substance which is a gaseous body is called *oxygen*, from the fact that it imparts to many of its compounds the character of an acid. This element combines readily with the common metals, transforming them into perfectly new and unrecognizable substances. Compare, for instance, a clean, bright

* The substance of this chapter has been freely drawn and condensed from the work of the celebrated German mineralogist, Francis von Kobell.

steel blade with a badly rusted one. In the former we see a dazzling metallic luster; and in the latter a dull red, or yellow friable earthy substance; and the cause of this wonderful change is nothing more than the combination of the iron or steel with this gaseous body, oxygen. Rust of a similar kind is formed upon all the *common metals,* while the *noble* ones resist for ages this enemy of their glory.

Chemists, who have appropriately been called the tyrants of the elements, may, it is true, unite by force, as it were, the noble metals with oxygen, nitrogen or hydrogen, three gaseous elements which exist in air and water; but their union is of short duration; the slightest touch or blow is sufficient to cause a frightful revolution, and the compound explodes with fearful violence, shattering everything with which it comes in contact. These combinations are better known under the name of the Fulminates of the noble metals.

Gold and platinum resist the action of almost every solvent that can be applied, only one substance, in fact, being known that effects complete solution, namely, chlorine, which is most readily obtained for this purpose by mixing one volume of hydrochloric acid with two of nitric. From this property the compound of these two acids has received the name of *aqua regia,* or king's-water. In consequence of their great resistance to foreign influences, the noble metals are generally found in a pure—that is, uncombined—or metallic condition; while, directly, the contrary is the case with the common metals.

Penetrating further into the analysis of the peculiarities of the noble metals, the great ductility which characterizes the king of metals, gold, beyond any other known substance, deserves our especial attention.

It is the glory of the sun that it possesses the power to spread and unfold its rays so widely, that it illumines not only the gorgeous palace of the sovereign, but also shines with a friendly light into the poor man's hovel, or silently creeps through the narrow chinks and crevices of the dreary dungeon. This comparison seems almost exaggerated, but the extraordinary ductility of gold admits of a diffusion which allows even the poorest to rejoice in its bright luster and sunny color. There are, doubtless, thousands who have never seen a ruby or sapphire, but none who do not know what gold looks like. When we reflect upon its wide diffusion in gilding and painting on innumerable familiar objects; on rings, pins, thimbles; on chains and watches, cups and vases, and thousands of articles of crockery-ware, we at once see that it has long been the common property of the world. Everyone knows the saying, that "a horse and its rider may be gilt with a ducat," but the divisibility of gold is far greater than this. A single grain of gold, not larger than the head of a very small pin, may be drawn out into *five hundred feet* of wire. Yet even this is not its utmost limit. If silvered wire is gilt and then rolled, a film of gold is produced not *one twelve-millionth* part of an inch in thickness.

The extraordinary ductility of gold was already known to the ancients, and, as Pliny relates, they were acquainted with the art of beating it and gilding on stone.

Silver is also exceedingly ductile, though not to such an extent as gold; platinum is inferior to silver.

The noble metals are likewise distinguished by being much heavier than the common ones, gold and platinum being more than nineteen times heavier than water, bulk for bulk.

The great weight of gold is turned to particular advantage in mining; clay and sand being hardly three times heavier than water, may be washed off, while the gold sinks to the bottom.

This is the basis of the process known as "panning."

The noble metals, moreover, are not volatile, and, generally speaking, unalterable by exposure to heat, while the common metals are all volatilized at high temperatures, or, if the air has access, combine with its oxygen. Taking all these characteristics together, we perceive at once that the noble or precious metals are gifted with pre-eminence in many respects, and while very generally distributed, they never run in large quantities, they have in this way acquired a still more exalted value.

When we ask as to the antiquity of these nobilities, gold and silver, at least, shine in this respect; for they are the two metals which have been longest known to man. A piece of gold and silver, in the pure state in which it always occurs, could not have remained long undiscovered, and its malleability and ductility would have insured its application to many useful purposes.

Silver is also of the highest antiquity, and a weighed quantity of the same was used as a trading medium already in the time of Abraham, and both metals are frequently mentioned in the history of the ancient

Egyptians, Phœnicians, Indians, Chinese and Scandinavians with the highest respect and veneration. The ancient Greeks made their drinking-cups of gold. In the "Songs of Orpheus," the wrestler, Anchaeus, wins a golden drinking-cup; Hercules gains a silver pitcher as a boxer, and Castor a golden horse for his skill in riding, etc., etc. In ancient history the metals are frequently compared to the planets. Hence, gold received the name of the Sun; silver, that of the Moon; and, among the common metals, lead bore the sign of Saturn; tin, of Jupiter; copper, of Venus; iron, of Mars; and quicksilver, that of Mercury, etc., etc. It is a well-known fact, however, that the value of the noble metals has increased immensely during the past few centuries; and since the Seeress of Prevorst has proved that, in a magnetic condition, she had an ungovernable inclination for gold, "the avarice of the miser who digs with a fiendish pleasure in his heap of coin, is, to a certain extent, justified." If the law were not so exceedingly limited in its views, and took notice of such experiments as that of the Seeress, the thief who stole a gold snuff-box would not be punished by far as severely as he who stole a brass one, and he would soon learn to explain that, at the time of the theft, he was in a magnetic condition.

The great value of gold and the influence it has ever possessed in governing the luxuries and creed of a people, naturally stimulated the fancy and gave origin to the endeavors which have been made from the earliest antiquity to produce it artificially. Alchemy, as a branch of the history of the precious metals, is of such absorbing interest, that I have thought it not out of the way to devote a part of the chapter to an account of this wonderful mania. The exciting task of the followers of alchemy was to convert the base metals, such as lead, copper or tin, into gold. Mystery and miracle would naturally form part of the means to obtain such ends, and the imagination wandered over a region which it was unwilling to leave. Still we cannot say that the transmutation of the metals is an impossibility; indeed there are many indications which render it highly probable, and many of the most eminent chemists, even of the present time, believe that the great dream of the alchemists will once be realized.

The first traces of alchemy, or the hermetic or spagiric art (from σπαειν, to separate, and ἀγειρειν, to unite), appear to be of Egyptian origin; and a fabulous Hermes Trismegistos is said to have founded it 2000 years before the birth of Christ; but it is only from the fourth century of our era that we begin to have distinct accounts of it.

The art of gold-making came from the Egyptians to the Greeks and Alexandrians, and subsequently to the Arabs; in the thirteenth century it had diffused itself all over Spain, France, England and Germany, and in 1700 it was raging as a mania all over the civilized world. The most important point in the alchemical creed was, that there existed a substance having the power of converting the base metals into gold or silver; this substance was called the "philosopher's stone," the "great elixir" or "magisterium;" it not only had the power of this transmutation, but it possessed the splendid property of making the old young again and prolonging life to an indefinite period. Thus Solomon

Trismosin, in a treatise entitled "Aureum Vellus," (1490) tells us (oh joy!) that it is an easy matter for him to keep himself alive by the use of this stone, and that he intends to live long enough to see "the last day."*

The process of obtaining the gold by the use of this stone was very simple; the base metal was only fused and small portions of the "great elixir" were then thrown upon it, when lo! a lump of pure gold or silver, as the case might be, was found in the place of the common metal.

The stone had wonderful power; one part being sufficient to "transmute" at least one million (1,000,000) parts of the base metal into gold. But how was this wonderful stone obtained? how could it be manufactured? "Ay, there's the rub!" The *older* alchemists believed that supernatural means were required for its production; some believed that the devil and other lesser demons and gnomes held the secret, which could only be purchased from them at the expense of the soul; others believed the secret was first made known to man by angels from heaven.

The alchemists of later times depended more on earthly means for producing it, and accordingly boiled, fused and distilled together the most diverse and wonderful substances their feverish imaginations could conjure up. Believing that many of my readers would like to learn the most bewitching art of gold-making, and as *we* firmly believe that this grand result can only be accomplished by *supernatural* means, we give here "a little recipe" by which they can conjure

* Where is Solomon?

up the foreman of the devils (for consultation only). This recipe was taken from an old work entitled the "Gold-Maker's Guide."

RECIPE.

Take of the gall of a black tomcat, killed when the night approacheth, one part; of the brains of a night owl, taken from out its head when the morning dawneth, five parts; mix in the hoof of an ass, when the tide turneth, and leave it until it doth breed maggots; place it upon thy breast-bone when the moon shineth bright—and—thou wilt see a sight which the eye of mortal man ne'er beheld afore.

Reader, should you not be successful, we can but encourage you by repeating the well-known nursery rhyme—

If you don't at first succeed,
Try, try again.

But we must not suppose that the alchemist was allowed to grope in the dark; by no means; many "guides" to the fabrication of the great stone were published, but unfortunately those that were written plainly never yielded an available stone, while those that were written mysteriously (as nearly all were) were, of course, not understood. The titles of the following works already give us an idea of the contents of such "guides." In 1649 appeared, "A Master-Key to the Opened Heart of Fatherly Philosophy," and the "Childbed of the Philosopher's Stone."

In 1700, "Philosophical Field Sports and Nymph-Catching," and the "Brightly Shining Sun in the Alchemical Firmament of the German Horizon." "Chymical Moonshine;" (Frankfort, 1744) the "Chymical King in his Robe of Purple," (1725) etc., etc. Many of these "guides" were sold very cheap, considering the immense wealth they contained, few costing more than a few cents.

Most of these "valuable guides" had very peculiar prefatory notices, for instance, "If thou be'st not altogether too stupid, and wishest not to climb *too high*, this little book will tell you how;" or, "Ye who do not seek too much but expecteth to find little, will be satisfied."

We would, however, wrong the alchemical writers if we were to attribute this mysterious manner altogether to an intent of deception; this was not always the case, it arose partly from the belief that it was wicked to make public this great secret, or that it might even cause the death of the writer before the pages were completed. Hence, Wilhelm von Schroeder, a distinguished chemist (1684), informs us, in a work entitled "Necessary Instructions in the Art of Gold-Making," that when philosophers speak *openly*, a deceit lies behind their words; while when they speak *enigmatically*, they may be depended upon.

This maxim seems to be very generally understood and practiced even by philosophers of the present day. In full descriptions of gold-making many preparatory notices are mentioned as absolutely indispensable. Thus: Reader, prepare first a "*philosophic mercury*" and a "*philosophic stone*;" these are thoroughly triturated together and exposed to a very gentle heat in a vessel of particular form; after a certain time they yield a black mass which was called "raven's-head;" by still continuing the heat, the substance becomes gradually white, and is then called the "white swan;" if the heat be now increased, and still continued for a certain time, the white mass becomes orange yellow and at last bright red; the operation is then completed and the stone obtained in its highest perfection.

First of all, a *philosophic quicksilver* was necessary, but where this wonderful substance was to be found nobody knew; some sought for it in the common metals; some in dew, rain, and snow; others (as the recipe given above proves) expected to find it in toads, snails, lizzards, snakes, owls, and plants and animals of all kinds. As there were so many "guides," the precious metals themselves could not be lacking. Thus the Danish ducats of 1647 were made of gold, obtained by artificial means by the alchemist of Christian IV., named Casper Harbach. So, under the Emperor Ferdinand III., 1648, a large medal was struck from gold that had been obtained in the emperor's presence by the transmutation of quicksilver; this transmutation being affected by means of a dark red powder which had been presented to the emperor by an unknown person. In like manner, the ducats struck under Landgrave Ernest Lewis of Hesse Darmstadt, were of artificial gold, produced, as was said, by the transmutation of lead; and the specie dollars of 1717 were of such silver.

In the year 1423, King Henry VI. of England issued several proclamations encouraging the study of gold-making, in order to obtain means to pay the debt of the state. Edward IV. of England, in 1476, accorded to a company "a four yeare privilege of making gold from quicksilver." The Elector Augustus of Saxony, who lived about 1560, was a great follower of alchemy, and had an extensive laboratory of his own which the people called "the gold-house," whether from the amount *made* in it or *spent* on it does not appear.

Kings, emperors and nobles (especially men who had plenty of money to spend) were always found at the head of the most ardent alchemical movements. In these times of wild turmoil and excitement, in which every *peasant* had his laboratory for "gold-making," there were still *a few* who kept cool. Thus Pope Leo X. once received a dedication of an enthusiastic poem, by an alchemist named Augurelli, in recognition of which His Holiness returned an empty purse with a note, that the happy man who was master of such pleasing art, was only in want of a purse in which to store the treasure he made.

Although it is true that there are many cases of gold-making in which no deceit was discovered, still there were many more which were openly proved to be false. The fate of those who were convicted of deceit, was to be hung up in a dress covered with tinsel; while those who "produced" gold, in a seemingly miraculous manner, often had as bad or even worse fate; they were seized, shut up, and ordered to make gold, to a certain amount daily, in default of which, they were beaten and kicked about in a most horrible manner, and, in fact, maltreated in every conceivable way. This fell to the lot of an unfortunate little tailor of Strassburg, who by some unknown means had obtained (from the devil he thought) a small fragment of the "great magisterium," but was entirely ignorant of its composition or manufacture.

It would be far beyond the limits of this little book to follow further the most interesting mania which raged for more than one thousand years, to observe how self-interest, passion, stupidity and madness everywhere followed in its track until the *science* of chemistry grew stronger and stronger, and one by one forced the old superstitious beliefs from the ground, and made way for new facts and great truths in the science of God's great creations.

Gold and silver are generally found in the metallic or uncombined condition. The form in which native gold occurs is rarely a regular one. It is mostly found in small granules, threads, dust or powder, spangles or laminæ.

It is softer than copper and fusible without difficulty. The extraction of gold from its ores is simply affected by crushing the ore to a fine powder and "panning," as above described, or by amalgamation. The process known as amalgamation, consists in treating the powdered ore with mercury, which combines with the metallic substances forming an amalgam or "butter;" this butter is placed in bags of buckskin and subjected to a strong pressure, the mercury oozes through, and a semi-solid mass remains in the bag, which is placed in iron retorts and distilled; by this process the volatile mercury is driven off and recondensed, and the gold or silver remains as a hard cake in the retort.

Gold is very generally diffused over the earth's surface, the largest quantities being found in the United States.

Germany, Russia, Australia, Mexico and South

America also yield large quantities of gold and silver annually.

Gold, as it is found in nature, is rarely without an admixture of silver.

The preparation of gold known as the purple of Cassius, or gold purple, consists simply of metallic gold in a very fine state of division, and is of great importance in the coloring and painting of glass and porcelain, imparting to it that magnificent purple color so much admired. This preparation of gold was first described by Andreus Cassius, in 1685, after whom it received its name, and it is formed most readily as a purple precipitate when solutions of the chlorides of gold and tin are mixed; but the preparation of a good article requires many precautions.

ALLOYS OF GOLD.

Gold readily unites with other metals by fusion, but its alloys with copper and silver are alone of importance. These metals, when added in fixed proportions, give gold a greater hardness, making it more fit for coinage and goldsmith's work.

The proportions of gold in an alloy is expressed in *carats*. A mark is divided into twenty-four carats; and if twenty-four carats of the gold alloy, for instance, contain ten parts of the foreign metal, it is said to be fourteen carats fine; if it contains two parts of the base metal, it is twenty-two carats fine, etc., etc.

Alloying with *copper* is called *red* carating; with *silver*, *white* carating.

SALTS OF GOLD.

There is but one important salt of gold, the perchloride, ($AuCl_3$) consisting of one equivalent of gold united to three of chlorine.

Besides this, there is the oxide of gold (AuO), the sulphide (AuS, AuS_3), the protocyanide (Au, Cy) and the percyanide (Au, Cy_3); but these compounds are of no interest whatever to the photographer.

SILVER.

Silver is found native in almost precisely the same form as gold. It is sometimes found in combination with sulphur in the form of black powder—sulphuret or sulphide of silver, known as argentite in mineralogy; and with chlorine as chloride or horn silver—it is sometimes, though rarely, found in combination with iodine as iodide, and with bromine as the bromide of silver. The smelting of silver ores is also very simple, and the extraction is only connected with difficulties when the silver ore is mingled in very small quantities through quartz or other rock. Silver is also found very generally distributed over the earth, the principal mines being in Mexico, Peru and other South American States.

ALLOYS OF SILVER.

By the addition of small portions of copper to silver the metal is rendered harder and more sonorous, while its color is not impaired.

All silver coin contains a certain amount of copper to render it more durable. The standard silver of this country consists of 92·5 parts of silver and 7·5 of copper. Besides this, silver forms various alloys with tin, lead, zinc, etc.; but these are of no practical importance.

SALTS OF SILVER

The principal salts of silver are the nitrate (AgO, NO_5), the chloride (Ag, Cl), the sulphide (Ag, S), the hyposulphite (AgO, S_2O_2, 5HO), the cyanide (Ag, Cy), the iodide (AgI) and the bromide (Ag, Br); these compounds being too well-known to every practical photographer to require any further description.

CHAPTER II.

THE FURNACE AND THE MANAGEMENT OF HEAT.

The subject, which necessarily is of primary importance, and has the first claim to our attention, is a suitable apparatus or furnace, in which the heat, in dry reductions, may be quickly and efficiently produced.

The imperfect knowledge of this important subject has been, as it were, the great stumbling-block of the practical photographer in attempting the reduction of his wastes. Nevertheless, it is a matter of great simplicity. Some persons have an idea that a furnace for the reduction of gold and silver residues involves a large outlay, much time and scientific knowledge; but this is an erroneous impression, which we shall try to refute in the course of this work. We shall, therefore, proceed to describe several simple devices, which are very efficient forms of apparatus for the smelting of gold and silver residues, and which can all be put together at the outlay of a merely nominal sum of money.

The main parts of every furnace are, first, the body in which the heat is generated; secondly, the grate or bars, upon which the fuel rests; thirdly, the ash-pan for receiving the clinkers and residue; and fourthly, the smoke-pipe for carrying off the gaseous products of combustion.

There are three different furnaces in use for the reducing operations. The *Wind Furnace*, in which the draft of the chimney alone urges the fire; common stoves belong to this class. If properly built, and provided with a high, clean and unobstructed chimney, they answer tolerably well for reducing operations. The *Reverberatory Furnace*, in which the flame of the fire is condensed and reflected upon the fused mass by a funnel-shaped dome, which will be more fully explained under another head. This furnace is best adapted for the reduction of gold and silver wastes, giving great intensity of heat with continuity of action; and lastly, the *Blast Furnace*, used to produce a quick and intense heat. The combustion, in this case, is urged by a current of air, forced through the fire by means of a bellows or rotating-fan. This style of furnace is used mostly in the refining or granulation of the precious metals. An ordinary small cylinder stove may be readily altered into the three different forms above mentioned. When used as a wind furnace, the degree of heat which can be produced in it depends upon the height of the chimney into which the flue passes, and, as a general rule, the higher the chimney the greater will be the heat.

It is, therefore, preferable, if possible, to have the stove placed in a cellar, or, at least, on the lower floor of a building. The intensity of the heat may also be vastly increased, by so proportioning the dimensions of the furnace and the chimney, that their diameters are equal, and the height of the chimney be, if possible, fifty or sixty times the diameter of the body of the furnace.

By following these directions, a wind furnace of the best kind can be obtained, but it is always preferable to use a reverberatory or blast furnace, especially in the reduction of paper ashes and sulphide of silver; the first being difficult of fusion on account of the large amount of silicious and carbonacious matter which it contains; the second being one of the most obstinate compounds, in regard to perfect reduction, known in metallurgy.

To construct a furnace on the reverberatory principle, the body should first be lined with a thick coating of refractory clay, that it may more perfectly withstand the intense heat to which it will be subjected.

This done, the orifice on the back of the stove, intended to receive the stove-pipe, should be plastered up with a thick lump of fire-clay. A funnel-shaped dome with an opening at the top, and large enough in diameter to fit tightly, as a cover on the top of the stove, should be ordered at the tinsmith's.

Figure 1 represents the dome with the smoke-pipe attached. The pipe should be self-supported, so that

Fig 1.

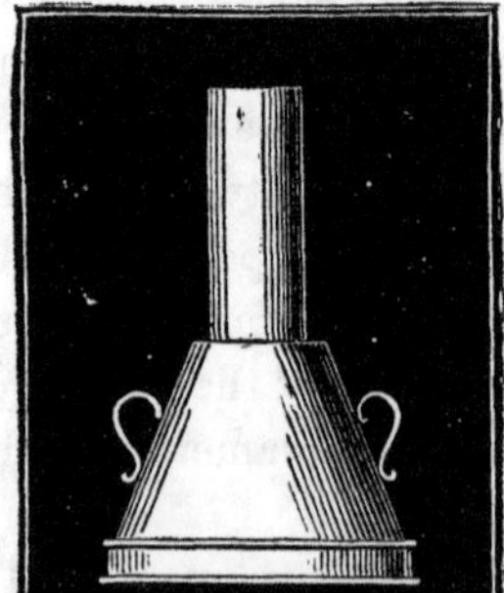

the dome may be lifted on or off to supply fuel, or to observe the progress of the operation.

Figure 2 represents this arrangement complete. A is the cylinder stove, of which the smoke-pipe B is plugged up with fire-clay; C is the funnel-shaped,

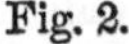
Fig. 2.

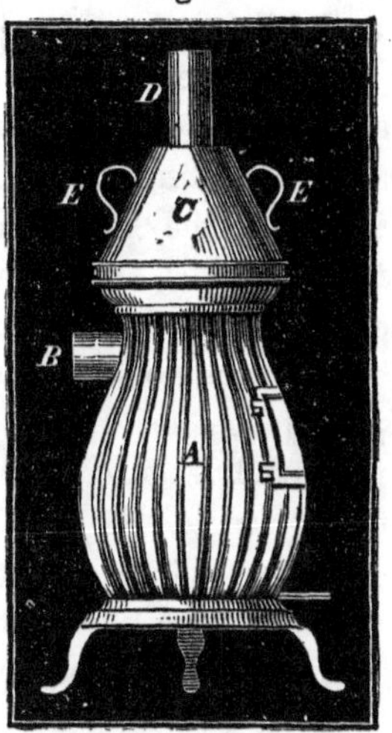

sheet-iron dome, tapering off into the smoke-pipe D; E E are two handles which may be attached to the dome to facilitate its removal. The great simplicity of this arrangement can be seen at a glance.

A Blast Furnace is constructed by setting a cylinder stove, with smoke-pipe attached, in the ordinary manner, closing the draft hole below the grate with a piece of sheet-iron, having a circular hole large enough to admit the nozzle of a pair of bellows, or rotating-fan arrangement. Of these two methods of creating a current of wind, we decidedly prefer the latter.

A very simple and efficient rotating-fan arrangement may be made as follows: Take a ribbon-block, two inches in diameter, and fasten into this at right angles four sheets of iron or stiff pasteboard, four inches wide by six long, or of any size convenient to the operator. A hole should be drilled through the center of the block, and a tightly fitting wooden axle passed through it. This arrangement completed should be mounted in an airtight wooden box, just barely large enough to admit of its turning freely, and a crank attached on the outside of the box to one end of the spindle. Two circular holes, about three-quarters of an inch in diameter, are now cut at both ends of the box, one an inch from the bottom, and the other an inch from the top; a piece of iron gas-pipe, about six inches long, being firmly secured in the lower hole, the arrangement is complete.

Figure 3 represents a sectional view of this wind arrangement. A is the ribbon-block; B B B B the four sheet-iron fans; C the spindle upon which the whole

Fig. 3.

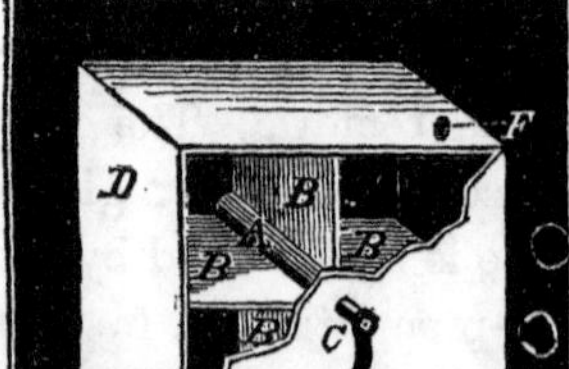

revolves; D the wooden encasement, into which the fan arrangement closely fits; E represents the crank, allowed to pass through the wooden side of the encasement, and by which the fans are set in rotary motion; F the circular opening at the back, one inch from the top of the box; G the orifice, with a pipe attached, through which the air is forced when the wheel is in motion.

When the fans are rapidly turned by the crank a strong and steady current of air is forced through the outlet. When it is required to use the blast, the outlet tube of the fan arrangement is allowed to project through the circular hole in the sheet-iron, placed before the draft-hole of the circular stove, and the connections are made perfectly airtight by a lute of fire-clay or putty. If the fan is now set in motion the air is forced through the fire, causing it to burn with the greatest intensity. The power of this little device is wonderful; and we have frequently brought a mass of gold, weighing twenty ounces, to thorough fusion in five to seven minutes.

When a cylinder stove cannot be readily obtained, a very cheap and efficient furnace may be put together as follows: Take a piece of stove-pipe, eight to ten inches in diameter and ten to twelve inches in height, and line it evenly from top to bottom, to the thickness of about one inch with fire-clay,* or, in default of this, with a mixture of equal parts of kaolin or pipe-clay and fine white sand.

This mixture should be made into an even, smooth dough or batter with water. Previous to the lining, however, a circular hole should be cut large enough to

* Fire-clay of the consistency of thick dough may be obtained at any fire-brick factory.

admit the nozzle of the rotating-fan arrangement, and about one inch from the bottom; or it might be provided with a small door to admit the draft, if used as a wind or reverberatory furnace. This being done, a series of short, stout pieces of wire, from one-eighth to one-fourth of an inch in thickness, should be embedded in the fire-clay, about one inch from the draft-hole, to serve as a grate. For convenience' sake, three sheet-iron legs should be riveted to the body of the furnace, so as to raise it six to eight inches from the ground. The bottom is provided with a movable sheet-iron cap, like those on blacking-boxes, to render the removal of the ashes more easy. The top being furnished with a movable sheet-iron dome and smoke-pipe, communicating with the chimney, the furnace is complete.

This simple little device is admirably adapted to the wants of the practical photographer, being small, neat, compact and very efficient. Used in connection with the wind arrangement, it is capable of melting steel in from fifteen to twenty minutes. In many cases, however, it may be more convenient to buy a complete reducing furnace. The neatest and most economical of the ready made furnaces, is that known as Kent's, which has all the appliances, so that it may easily be changed into either wind, blast or reverberatory furnace.

SELECTION AND PREPARATION OF FUEL.

Coal, coke or charcoal is the fuel most generally used in furnace operations. Coal is the least available as it yields a large amount of ash and clinker, which choke the grating and seriously impede the draft. Coke or charcoal are, therefore, decidedly to be preferred in furnace operations. Weight for weight, their amount of heat is almost equal; but the much greater density of coke enables it, bulk for bulk, to give a greater heat by ten per cent. For all operations requiring a high heat, we prefer to use coke, as, when it is of good quality, it yields but little clinker and ash, and gives an intense and steady heat. Whether coke, coal or charcoal be used, they should be broken into pieces of the size of a large walnut, so that they may not pack too tightly nor too loosely around the crucible when in the fire.

A *mixed* fuel has many advantages over any *single* one of the above-named fuels. The following will be found among the best: Coke, one bushel; charcoal, one peck. Or coke, one bushel; charcoal, one peck; hard coal, one-half peck. Or charcoal, one bushel; coke, one bushel; and coal, one-half bushel. These materials should be previously pounded to the proper size and thoroughly mixed. They are considerably improved, if they are to be used without blast, by being moistened with water, containing one ounce of niter to the gallon, and then dried thoroughly.

CRUCIBLES AND MELTING-POT.

The vessels or receptacles used in fusing, or smelting substances, are called crucibles. The Hessian or French crucibles are those most generally used. The Hessian, so called from their place of manufacture, are of a tapering form, either round or triangular, and come in nests, gradually increasing in size, from one-half ounce upward. They are made of a very refractory and durable material, and, being very cheap, are preferred to all others. The Plumbago or Black Lead crucibles are suitable for refining only, and should not be used in reducing operations. The eight or ten ounce sand crucibles are those most convenient; being much thinner than the larger ones, the reduction may be performed in a much shorter time, and the cost of these is but a few cents. The French crucibles have but one fault, namely, their high price, which excludes them entirely from common use.

We close this chapter by giving a list of what is required for a complete furnace outfit, the whole of which will not cost more than five dollars.

A small furnace or cylinder stove, fitted up as above described.

A fan blast arrangement, or a pair of double acting bellows.

One or two pair of iron tongs, *stout* but not too heavy.

Several Hessian sand crucibles with covers.

Several lengths of clean, stout iron wire with which to stir the fused mass.

Some fire-clay, or putty, to be used as lute, in the blast arrangement.

CHAPTER III.

ON THE BEST METHODS AND APPARATUS FOR SAVING THE VARIOUS WASTES OCCURRING IN PHOTOGRAPHY.

We shall in this chapter describe the methods employed in saving the various wastes and residues, in the order in which they are produced, in the different photographic manipulations. We will begin with the preparation of the bath for sensitizing the plate, and proceed with the description to the toned and finished print, with a view to the greatest economy and saving of the precious metal, in all the necessary operations.

THE BATH.

In the preparation of the bath, it is customary to saturate the solution of nitrate of silver with an iodide or a bromide to prevent the blemishes upon the plate, known as pinholes. The canary yellow precipitate, produced by the addition of one of these reagents, consists of pure iodide or bromide of silver, which should be carefully saved. It is best removed from the solution by filtration through paper. The paper used in this operation, of course, becomes saturated with the solution, and acquires a large percentage of silver.

The filter, with its contents, should be well dried, and carefully preserved in a covered wooden box.

THE SENSITIZED PLATE.

The sensitized plate, before being placed into the holder, should be thoroughly wiped on the back from all adhering solution; a pad or cushion may be placed near the bath for this special purpose, and a large saving attained by this simple device. A roll or pad of blotting-paper may be advantageously substituted for the cotton. By simply drawing the back of the plates across it, when removed from the bath, the cushion becomes saturated with crystals of nitrate of silver in a short time.

A clean, porous, cotton cloth should be placed, or hung, in a convenient proximity to the plateholder, the corners of which should be carefully and thoroughly wiped dry from adhering solution after each exposure. This operation should *never* be omitted, as better photographic results, as well as greater economy, are thus obtained. Whenever a sensitized plate, either before or after development, becomes useless from any cause, the collodion film should be carefully removed from the plate, and thrown into a bottle containing salt water. If it is desired to reduce these films by the wet method, in preference to fusion, the undeveloped and developed should be kept apart, otherwise they may be mixed.

DEVELOPER RESIDUES.

When developing a plate, all superfluous solution from the same should be allowed to drain into a bottle, kept separately for this purpose. When the development is finished, the plate should be flowed with a small amount of water, which is allowed to drain into the same bottle; the plate may then be thoroughly washed under the tap. In large galleries, where a considerable number of plates are developed, a small, watertight tank of sheet-iron, or cemented wood, may be substituted for the bottle.

THE FIXING SOLUTION.

The print fixing solution, whether hyposulphite of soda, or cyanide of potassa, should, when too weak for fixing purposes, be placed in a stoppered bottle and labeled accordingly.

When removing a plate from the fixing bath, it should be held for some time in an upright position, so that all superfluous solution may trickle back and be retained. If, after fixing, the plate is accidentally scratched, or otherwise rendered unfit for use, the film may be scraped off and preserved in a separate bottle.

OLD NEGATIVES, AMBROTYPES, ETC., ETC.

When it is desired to again use the plates upon which negatives or positives have been taken, the silver contained in the film of the picture may also be regained, to do which proceed as follows:

Half fill a deep earthen dish with the following solution:

Caustic potash - - - -	8 ounces.
Water - - - - -	16 "

OR,

Carbonate of potash - - -	10 ounces.
Boiling water - - - -	15 "

Into either of these solutions, preferably the first, place, face downward, a number of negatives, one upon another, until the dish is full. The negatives are allowed to remain from one to four hours, after which time the varnish will be completely softened, and the whole film may be readily removed with the fingers, and placed *unwashed* into a clean labeled bottle.

The plates, if intended to be reused for the production of a picture, must, after this operation, be washed with dilute nitric acid, to remove all superfluous alkali, well washed, and finally rubbed dry with a clean cloth. By these means a brilliantly clean plate is obtained.

PAPER CLIPPINGS, ETC., ETC.

The filters and residues obtained in sensitizing paper, should be carefully saved. All untoned paper contains a large amount of silver, in the form of chloride, and furnishes one of the richest and most valuable residues obtained in the gallery. For the sake of economy, especially in larger establishments, the prints should be cut to the proper size, as they are removed from the frame, before toning, as thus a large amount of valuable clippings are obtained, the silver on which is otherwise lost in the toning and fixing solution, from which it cannot be so readily regained. Paper filters and filtering cotton are the richest of the paper wastes, containing generally from forty to sixty per cent. of silver, according to the number of times the silver solution has been passed through them. The drippings of the sensitized sheets of paper should be received upon sheets of newspaper, which may remain there until they become saturated by the solution.

THE PRINT WASHINGS.

On removing the prints from the frame, they should

be placed in a small tank of water and thoroughly shaken or agitated for a few minutes.

The milky liquid resulting, is poured into a clean barrel, and the operation repeated several times, or until the water ceases to show a milky turbidity. In each case the smallest possible amount of water should be used, say one gallon to each hundred pints. Three washings will generally suffice, and the prints may then be thoroughly washed in abundance of water, without any further loss of silver. About a pint of the following solution should be added to the washings of each hundred prints.

Water - - - - - - - 1 pint.
Common salt, to thorough saturation.

The barrel and its contents should be thoroughly stirred or shaken after each addition of the salt water.

The salt solution should *not* be added to the water, in which the prints are washed, as is customary with some, as it only tends to form the insoluble chloride within the fibers of the paper and decreases the amount of the soluble silver; but, on the contrary, the purest soft water should be selected for the purpose. Care should be taken that no waste hyposulphite or cyanide finds its way into the barrel containing the print washings.

THE TONING SOLUTION.

When the toning solution becomes unfit for use, it should be poured into a large stoppered bottle and carefully preserved. It contains a small amount of gold and silver. We once obtained twenty dollars' worth of pure gold from the six months' saving of a gallery. The toning bath residue may be treated equally well, either by the dry or wet methods of reduction, as will be hereafter described.

THE PRINT FIXING SOLUTION.

The print fixing solution, when it becomes weakened by use, contains a considerable amount of hyposulphite of silver. The prints should be well drained from all superfluous hyposulphite solution before they are removed to the washing-tank.

It is customary in some galleries to add fresh hyposulphite of soda, as the old solution becomes weakened in action; this is, however, false economy; the prints are never so perfect or lasting when this method is resorted to. The old solution should be poured into a barrel or demijohn, properly labeled for future treatment, when enough has accumulated.

THE APPARATUS.

The necessary apparatus to carry out the operations described in this chapter is very simple. Several funnels, and earthen dishes, a few large, clean bottles, or demijohns, and a ten or twenty gallon barrel, constitute the whole outfit. As a tank for containing the developer drainings, a tightly dove-tailed, shallow wooden box is procured, the sides and bottom of which are evenly covered with a thick varnish like composition, impervious to water and made as follows:

Best yellow resin - - -	16 ounces.
Yellow beeswax - - - -	12 "
Turpentine - - - -	4 "
Lamp-black, enough to color.	

The wax and resin are melted together in a tin pot, and, when perfectly liquid, the turpentine is stirred in. The mixture must be applied hot. The barrel containing the print washings should be provided with a stopcock or faucet, placed about twelve inches, instead of two inches, from the bottom of the barrel. When the curdy precipitate has settled to the bottom (for which the liquid should remain at rest for at least twenty-four hours), the sides are gently thumped all around with a mallet or hammer, so that the precipitate, clinging to the sides of the barrel, is disengaged. The faucet may then be turned on and all the superfluous water allowed to flow out. This operation can be continued until the precipitated chloride reaches the height of the stopcock.

A few glass syphons, or a yard of narrow rubber tube, should be provided, which form a convenient method of removing the water from above an easily disturbed precipitate.

For the solid wastes, several clean, wooden boxes, with tightly fitting covers, properly labeled, should be kept in a convenient place.

We would here enjoin upon the photographer to keep each residue, either liquid or solid, in a separate bottle or box. To avoid mistakes, these should be labeled respectively, viz. : Collodion Films, Developer Residues, etc., etc. Some photographers have a habit of mixing indiscriminately together the most diverse compounds obtained from the various residues ; this occasions much trouble and loss of time in after operations, and should be carefully avoided.

CHAPTER IV.

TREATMENT OF THE VARIOUS WASTES PREPARATORY TO SMELTING.

HAVING, as we will suppose, collected a considerable quantity of gold and silver residues, the next step is to prepare them for the crucible. For this purpose the solid wastes are reduced to ashes, and those of a liquid nature brought to a solid condition by precipitation.

BURNING PAPER WASTES.

A great deal depends upon the proper burning of the paper residues, and it is necessary that we enter a little more closely upon this subject. To insure a perfect combustion, it is necessary that the air has free access to the burning mass. The most convenient method to insure this result, is to burn the paper in a fireplace grate. The grate is filled loosely with the clippings, old filtering-paper, etc., etc., which is then ignited, and, when it has burned down to a glowing heap of ashes, the rest of the paper is fed very slowly by handfulls, so as not to suffocate the flame for an instant. The ashes should be carefully raked into an iron pan when they begin to choke the draft, and spread thinly, so as to allow them to glimmer on. By these means a very perfect combustion is insured. When well burnt, the ashes should have a whitish gray color, without any black or carbonaceous residue; should be friable and easily reduced to a fine powder by friction between the fingers. The richer the ashes in silver, the whiter and more metallic they will be in color. In the ashes of filtering-paper, large granules or spangles of metallic silver are sometimes found. Those who cannot easily obtain the use of an open grate, should burn the paper in a small cylinder stove, or in a clay or iron fire-bucket. In using a stove with a good draft, care should be taken that none of the ashes are whirled up the chimney, as a great loss can be occasioned thereby. Filters, filtering-cotton, cloth, untoned clippings and other paper residues may be all mixed together. It is not necessary to keep these separate, as practiced by some, except when the paper is sold to refiners or smelters. It is always advantageous when cloth or paper, poor in silver, are burned, to moisten them with either of the following solutions, which insures a much more perfect combustion:

Nitrate of potash (saltpeter) -	6 ounces.
Water - - - - - -	16 "

OR,

Chlorate of potash - - -	4 ounces.
Water - - - - - -	16 "

Of these two solutions the first is the cheapest, and both are equally efficient.

The paper, or cotton, as the case may be, should be sprinkled with one of the solutions, until thoroughly dampened through, and then perfectly dried by the heat of the sun or a gentle fire. When ignited in a stove, the paper thus prepared, be it ever so poor in silver, burns with intensity.

TREATMENT OF DEVELOPER WASHINGS.

The turbid liquid gathered from the plate washings is mixed with a strong solution of common salt, and exposed for several days to bright sunshine; at the end of this time the liquid will generally be found clear, and a black precipitate to have subsided to the bottom, which consists of a mixture of metallic silver and iron. The bottle should be well shaken, and the whole contents poured upon a paper filter and allowed to dry. Another method is to add to the "washings" a solution of sulphide of potash, until it ceases to give a black coloration in the liquid; it is then likewise thrown upon a filter and allowed to dry. The precipitates thus prepared are ready for the crucible. Of these two methods we prefer the first.

OLD COLLODION FILMS.

To prepare these in the best manner, they should be first thoroughly dried and then sprinkled with either of the solutions recommended, on the last page, for paper clippings, the last of which is preferable. After having thoroughly dried them, they are placed upon an iron shovel or pan and ignited with a glowing coal, when they will be reduced to a fine yellow ash. The most convenient method of separating the water, or adherent solution, from old films, is to throw them, when removed from the plate, upon a fine cloth filter, suspended in the neck of a wide-mouthed bottle. The solution flows through, and may be afterward precipitated with common salt, while the solid films are left upon the cloth to dry. The films obtained in the various stages of the process, should be kept apart. Thus, those obtained *before* development, contain a large amount of iodide and bromide of silver, as also considerable unused nitrate of silver; those *after* development, metallic silver, iron and nitrate of silver; while those obtained after *fixing*, or when reusing an old plate, contain *only* metallic silver, with a solution of hyposulphite of soda, or caustic potash, as the case may be. It would, therefore, be convenient to have four different bottles for the reception of the various films; each should be properly labeled, and in no case should they be mixed together.

THE HYPOSULPHITE FIXING BATHS.

When a sufficient quantity of this solution has accumulated, either from the prints or plates, it should be placed into an open barrel, a large wooden tank, or any other convenient vessel, and a solution of sulphide of potash added as long as it produces a black precipitate or coloration; the liquid is then allowed to remain at rest for a few days; and when the solution is perfectly clear, as much water as possible, without disturbing the precipitate, is drawn off. The muddy residue is poured upon a large filter, and the precipitate, after being thoroughly washed with warm water, is allowed to dry.

The black mass thus obtained, which consists of sulphide of silver, should be placed on an iron pan or shovel and exposed to a brisk heat over a lively coal fire. This operation must be conducted either in the open air or in a fireplace, as thick vapors of sulphur are generated during this process of ignition, or roasting, as it is termed.

When the black mass has been exposed to a red heat for a few minutes, and has fused into an even "glass," and all vapor ceased, the shovel or pan may be removed from the fire and its contents allowed to cool.

This operation may be dispensed with when the iron flux, as recommended in the next chapter, is made use of, or when the sulphide is obtained from the developer washings, as these already contain the amount of iron requisite. Pure sulphide of silver yields about half its weight of metallic silver.

TREATMENT OF THE CYANIDE FIXING SOLUTION.

It has been generally customary to throw away waste cyanide fixing solutions, as they became weak by use. They, contain, however, a large amount of silver, which may be readily and economically utilized. The precipitation of silver from a solution of its cyanide, depends, of course, on the addition of a reagent, having a stronger chemical affinity for the silver than the cyanic acid with which it is combined in the solution. This manipulation is attended with considerable danger, owing to the large amount of cyanic acid gas, a most potent poison, which is evolved during the operation. But with proper management and care, not only the silver, but the active principle of the cyanide of potash, may be regained from such waste solutions.

Take a wide-mouthed bottle, of about two quarts capacity, into the neck of which fit a smooth and very tight cork, through which pass two glass tubes, one reaching within a few lines of the bottom of the bottle and projecting an inch over the cork, the other passing only about half an inch *through* the cork and bent at right angles on the outside. The neck of a small funnel is secured to the top of the longest tube by means of a little wax. Figure 4 represents this arrangement. A second bottle which may be about half the size of the former, is provided with a tight cork and a glass

Fig. 4.

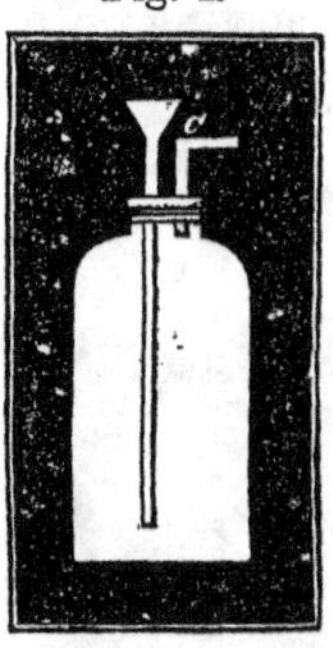

tube of the same diameter as the last (which should be about a quarter of an inch), is let through this until it reaches within the fourth of an inch of the bottom, and like the former is bent at right angles. Having completed this arrangement, half fill the first bottle with the waste cyanide solution, and into the second one place a solution consisting of—

Pure caustic potash - - -	4 ounces.
Water - - - - - -	1 "

And make the cork and joints of both bottles perfectly airtight by brushing over the following mixture:

Beeswax - - - - - -	2 ounces.
Resin - - - - - -	¼ ounce.
Turpentine - - - - -	½ "
Vermilion, enough to color.	

Place the wax and resin on the fire in a tin vessel,

and when thoroughly melted, stir in the turpentine and vermilion. The latter is only for appearance and may be omitted.

Figure 5 represents the complete arrangement. A is the larger bottle containing the waste cyanide solution; B the longer tube with the funnel attached; C the shorter tube connecting with the rubber-pipe D, through which the cyanic acid gas evolved is lead by means of the glass tube E into the solution of pure potash F.

Fig. 5.

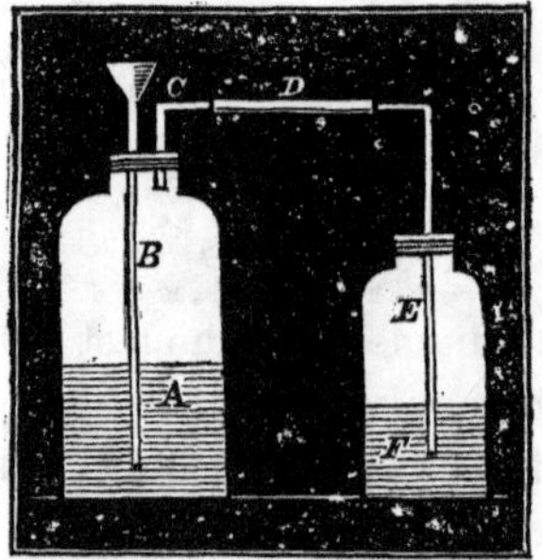

Having concluded these preliminary arrangements, common hydrochloric or muriatic acid should be slowly added to the cyanide solution in the larger bottle by means of the small funnel. Upon each addition of the acid the solution foams and seathes violently, and care should be taken that it does not boil over. Simultaneously with the addition of the acid, a white, curdy precipitate of chloride of silver is formed, which, however, is at first redissolved as it descends into the solution. Acid should be added as long as it occasions the slightest precipitate or foaming in the solution, or until it is considerably in the access. At the close of the operation a heavy precipitate of pure chloride of silver is found in the first bottle, and a solution of cyanide of potash in the other. If, after the close of the operation, the solution of cyanide of potash in the smaller bottle should be somewhat weak, another quantity of the waste cyanide solution may be treated with acid and the resulting gas allowed to pass into the weak liquid. It, however, generally happens that the fresh cyanide solution obtained by this method is too strong for photographic purposes; in that case it may be simply diluted with water. This process works very beautifully and the resulting substances are nearly chemically pure. When it is desired to save the silver only, the waste cyanide solution may be precipitated in an open vessel with strong hydrochloric acid as before mentioned; but the operation should be conducted in an open place with the back against a strong draft of wind, so that the poisonous fumes arising may be carried away from the operator. We would recommend, however, the first process as the best, having used it with great success in the reductions of the waste cyanide solutions, occurring in the electro-plating art.

Another very excellent method of utilizing the cyanide solutions, is to evaporate them to dryness. By these means a solid is obtained (containing all the silver), which may be employed with the greatest success as a flux for the reduction of *sulphide* of silver and other wastes. In evaporating the solution great care is also necessary, as fumes of prussic acid are given off. The evaporation should be conducted either in a fireplace or in the open air.

HOW TO PRECIPITATE THE TONING BATH.

At least three gallons of the waste baths should have accumulated before it is worth while to undertake their precipitation; then place it in a wide-mouthed glass vessel, and add the following solution in small quantities, until it ceases to give a black precipitate:

Sulphate of iron - - -	4 ounces.
Water - - - - -	16 "

Allow the now jet black solution to remain at rest for several days, or until the precipitate has entirely subsided to the bottom; then add a few drops of the iron solution to the clear liquid, and if no further precipitate is occasioned thereby, the operation is complete. Filter and dry the ebony black powder thus obtained, which consists of a mixture of metallic gold, iron and the oxide of iron, and expose it to a red heat in an iron shovel or other convenient vessel. By these means the whole of the iron is changed into the sesquioxide, thus rendering its removal very easy. When quite cold, place the now dark red powder in an evaporating-dish, and pour upon it some pure hydrochloric acid and heat the vessel gently. The solution of the iron immediately takes place, and when the seathing has ceased, dilute the mass with cold water, throw the contents of the dish upon a filter, and wash the remaining black powder copiously with water. The result is pure metallic gold. Care should be taken that the acid used in this operation is pure, or does not contain any uncombined gas, as this would cause a partial solution of the gold, and thus occasion loss. It is not necessary to remove the iron, except when the gold is to be used for the production of the chloride. If the product is to be fused, the precipitate obtained by the addition of the sulphate of iron, may be dried and fluxed without further preparation.

PREPARATION OF THE NITRIC ACID PLATE CLEANING SOLUTION.

When nitric acid has been used for removing old films from glass plates, it takes up a considerable amount of metallic silver; this may be regained in the form of chloride by adding some pure hydrochloric acid. When potash, or its carbonate is employed, as directed on page 41, for the removal of films, no silver is employed, but all is retained in the films.

PRECIPITATION OF OLD BATHS, NITRATE OF SILVER SOLUTIONS, ETC., ETC.

When the silver bath has become weak and uncertain in its action, or has been spoiled by accidental admixture of some foreign substance, the quickest and most expeditious method of utilizing it is to precipitate it. This is done by adding to such residues a strong solution of common salt, or some dilute hydrochloric acid. If the silver solution is pretty strong, it should previously be diluted with an equal bulk of water, and the acid or salt added as long as a white precipitate is caused by such addition. The result is pure chloride of silver. It should be thrown upon a filter and freely washed with warm water. Chloride of silver may be very efficiently reduced by the wet methods, as will be hereafter described; but for this purpose it must be freshly precipitated, or at least not have been allowed to dry. Chloride of silver which is to be reduced in the humid way, may be conveniently kept in the dark-room, in a bottle containing enough water to thoroughly cover it, until a sufficient quantity has accumulated. Another excellent method of treating waste solutions of nitrate of silver, which, however, is only applicable, if intended for humid reduction, is to precipitate by means of sulphuric acid or a sulphate. The following solution answers very well:

Sulphate of potash - - -	4 ounces.
Water - - - - -	12 "

Add it slowly, drop by drop, into the silver solution, as long as it produces a white precipitate or coloration. The white powder resulting, which consists of sulphate of silver, after being well washed, should be kept under water like the former. The sul*phate* of potash, used in this formula, must not be mistaken for the sul*phide*, which produces the black precipitate of sul*phide* of silver. From all waste nitrate of silver solutions, the silver should be precipitated in the form of chloride, and the sul*phide* avoided as much as possible, except in such cases where it is indispensable, as in waste hyposulphate solutions, etc., etc. *Chloride* of silver is the easiest, *sulphide* of silver the hardest and most obstinate compound to reduce.

GENERAL REMARKS AND PRACTICAL SUGGESTIONS.

Precipitation.

The process of precipitation, as we have already shown, is employed for the immediate separation of a body in a solid state from a chemical solution. As heat in all cases promotes the subsidence of the precipitate, the solution may be previously warmed and the reagent added slowly and under constant stirring, so that the parts of the liquid may be brought into contact. The solution is then allowed to remain at rest until the deposition of the precipitate has left the supernatant liquid clear. A few more drops of the reagent should now be added, and, if no further cloudiness or precipitate is occasioned, the operation is complete.

A liquid may be precipitated in any convenient vessel, but this should preferably be of glass, as then the action and result may be better observed. When but small quantities of a solution are operated upon, glass vessels with tapering sides, known as precipitating glasses (Figure 6), may be employed with advantage. The form of these vessels is peculiarly well adapted

Fig. 6.

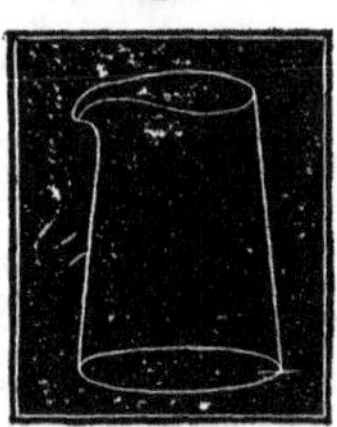

for the precipitation of precious salts, as none of the precipitate can adhere to the sides of the vessel. When a very large quantity of solution is to be precipitated, the operation is best conducted in large, wide-mouthed glass bottles or candy-jars, or in glazed earthen vessels, though the latter are open to objection, on account of not being transparent. An excellent vessel for precipitating solutions, is the barrel with the faucet in the middle, as recommended on page 45, for print washings. A very thorough washing of the precipitate is insured by this simple device.

Filtration and Decantation.

The mode most generally resorted to for the separation of precipitates from liquids in which they are suspended, is that of filtration. The process consists simply in passing the substance to be operated upon through some medium, fine enough to intercept any solid particles, however finely divided, still of sufficient porosity to allow water to pass through readily. Paper is the substance most generally used for this purpose. The paper best adapted for the filtration of the bulky and heavy gold and silver precipitates, is that known as German filtering paper, which, besides being very strong in texture, allows the water to run through very rapidly. When a precipitate is of a bulky nature, and, especially if accompanied by much water, a *plaited* filter should be made use of, as it prevents a close adhesion of the paper to the glass, thereby greatly expediting the process.

To form a plaited filter, take a square of good porous paper and fold it diagonally, as in Figure 7. Turn A upon B to obtain the crease E, and open it; then

Fig. 7.

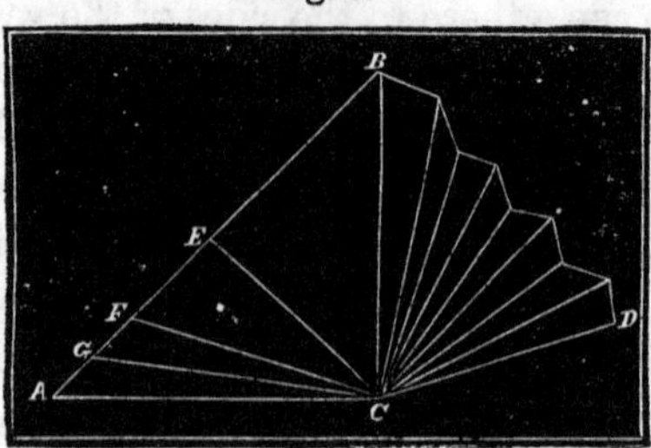

double A upon E in the same direction to make crease G; and holding this plait between the fingers, make the fold between F and D; divide the space between E B and B D in the same manner.—*Morfit.*

Figure 8 shows the position of the filter in the funnel. After the filter has been properly secured in the funnel, and previous to the addition of a liquid containing a precipitate, the filter should be thoroughly wet with some clean water, as by these means the pores of the paper are opened, rendering the filter less

Fig. 8.

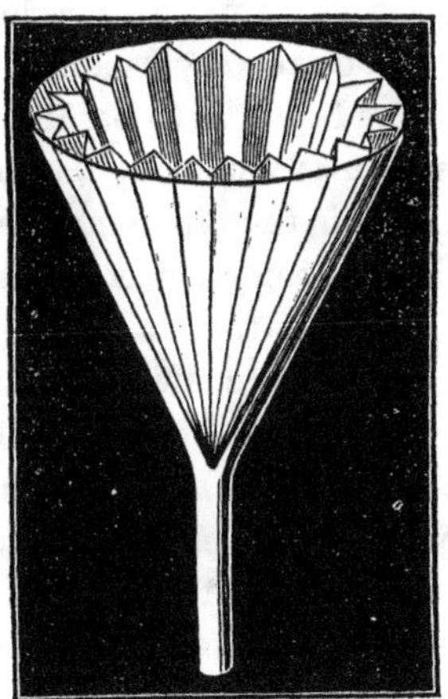

liable to clog than if the turbid liquid were added immediately. There are, however, some precipitates, like the chloride and sulphide of silver, which so completely clog certain kinds of paper, as to prevent the slightest passage of water. In such cases a *bag of felt* must be made use of.

A very convenient filter for chloride of silver, when

large quantities are operated upon, is an old felt hat, suspended from four corners by means of strings, which should be fastened to the ceiling, just above the sink. When paper is used for filtering liquids which pass through very slowly, a series of small funnels with plaited filters should be made use of. Large funnels are objectionable, as the filters are very liable to tear from the weight of the liquid, having no proper support at the bottom; but this trouble may, in a measure, be remedied by placing a piece of cotton loosely in the barrel of the funnel, so that it reaches to the paper, and it will now act as a support for the filter. Loose plugs of cotton alone are very useful.

When a corrosive liquid is to be filtered, as a strong acid or alkali, a plug of asbestos, placed loosely in the barrel of the funnel, is the only suitable medium.

Decantation has the same object as filtration, namely, the separation of water from a precipitate; it is only resorted to when the sediment is very heavy and subsides readily to the bottom of the vessel without being easily agitated, as, for instance, the chloride of silver. It consists simply in inclining the vessel gently to let the supernatant fluid run off, or in draining it off with a syphon. The syphon is a very convenient little instrument and should always be kept at hand.

The most simple form of a syphon is a small glass tube, bent so that it has two arms of unequal length. (Figure 9.) The long arm may be twenty inches in length, the shorter one about fifteen inches. The syphon is filled with water, the mouth of the longer arm closed with a finger, and the shorter branch introduced, mouth downward, into the liquid to be decanted, until it nearly reaches the level of the precipitate without disturbing it. Upon removing the finger

Fig. 9.

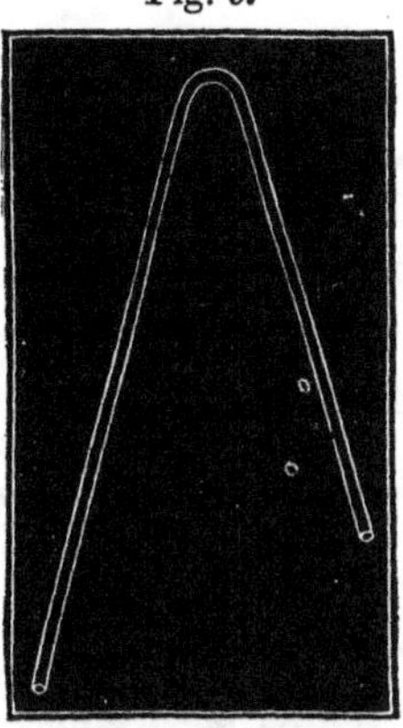

the liquid runs out in a continuous stream, and may be almost wholly drawn off. If the liquid is not injurious or unpleasant to the taste, the syphon may be inserted into the liquid without previous filling, being drawn over by suction with the mouth through the long end.—*Morfit.*

A small piece of rubber gas-tube makes an excellent syphon; it should be filled with water and placed into the liquid in such a manner, that the end hanging over the side of the vessel is considerably shorter than that in the liquid. All precipitates should have the superfluous water decanted off before they are thrown upon the filter, as much time is saved thereby.

Drying.

When much of the water contained in a precipitate has drained through the filter, and the precipitate is of the consistency of butter, it should be removed from the filter with a piece of glass, or a spatula, and spread on a large sheet of paper and allowed to dry spontaneously; or it is thrown upon an earthen dish and placed upon the stove until it is completely dry. Where there is no occasion to hurry up the drying, the precipitate may be allowed to remain upon the filter or exposed to the sun on a dish. When a large amount of chloride or sulphide of silver is to be dryed, it should be placed in a large earthen bowl, or porcelain evaporating dish, and exposed to the naked fire, under constant stirring, until it is perfectly desiccated.

Roasting.

Some substances, like the sulphide of silver and the precipitate of the toning bath, must be exposed to a low red heat. This operation is termed "roasting," or "ignition," and is resorted to either, as in the case of sulphide of silver, to partially expel a substance from a compound, or, as in the case of the toning bath, to so modify one of its constituents as to make it soluble. Most substances, when exposed to heat in this manner, give off disagreeable odors. The operation is therefore best conducted in the open air or in a fireplace. A small clay fire-bucket, filled with burning coal, should be used to give the necessary heat, and the substances exposed to it in a thin layer on a shovel or frying-pan. An exposure of fifteen minutes will generally suffice, if the fire is lively.

Evaporation.

The evaporation of caustic solutions, like the cyanide of potash, is best conducted in porcelain evaporating dishes. This operation should also be performed in the open air or well ventilated places. The solution should be allowed to boil rapidly, until all the water has disappeared. The dish is then removed from the fire, and the salt thus obtained placed in well stoppered bottles.

PRECIPITATING SOLUTIONS.

For Developer Washings.

Common salt - - - - 8 ounces.
Water - - - - - 20 "

OR,

Sulphide of potash - - - 4 ounces.
Water - - - - - 16 "

For Hyposulphite Fixing Baths.

Sulphide of potash - - - 8 ounces.
Water - - - - - 24 "

OR,

Sulphide of potash - - - 4 ounces.
Water - - - - - 16 "

For Toning Bath.

Sulphate of iron - - - 4 ounces.
Water - - - - - 16 "

OR,

Sulphate of iron - - - 4 ounces.
Water - - - - - 20 "
Sulphuric acid - - - - 4 drops.

OR,

Protochloride of tin - - - 1 ounce.
Water - - - - - 12 ounces.

The last solution answers very well, but is too expensive for general use.

The Nitric Acid Plate Solution.

Hydrochloric acid - - -	4 ounces.
Water - - - - -	8 "

OR,

Sulphuric acid - - - -	4 ounces.
Water - - - - -	10 "

The last solution should only be used when the precipitate is to be reduced in the humid way.

For Old Baths, Nitrate of Silver Solutions, etc., etc.

Common salt - - - -	8 ounces.
Water - - - - -	20 "

OR,

Hydrochloric acid - - -	4 ounces.
Water - - - - -	4 "

OR,

Sulphate of potash - - -	4 ounces.
Water - - - - -	12 "

The last should only be used when the result is to be reduced by the wet method.

CHAPTER V.

FLUXING AND FUSION.

The liquifaction of a substance by the application of heat, is termed fusion. Fluxes are substances usually of a saline character, which are mixed with a compound in order to accelerate its reduction or decomposition by heat. Chloride of silver, for instance, consists, as we have seen, of metallic silver, combined with the gaseous body called chlorine, with which it forms a white powder. The chlorine being very firmly combined with the silver, can only be separated from it by adding another substance for which it has a still greater attraction or affinity than for the silver. The substance usually added for this purpose is an alkali, or alkaline carbonate (soda or potash). When the chloride of silver is mixed with one of these substances and exposed to a red heat, the chlorine leaves the silver and combines with the alkali, forming the chloride of potash or soda, as the case may be, and liberating the silver in the uncombined or metallic condition. This is the rationale of reduction. The principal objects of fluxing are :

1. To cause the more ready fusion of a body, which it is difficult or impossible to fuse by itself.

2. To fuse a foreign substance mixed with a metal, in order to separate the latter by its difference of specific gravity.

3. To prevent the formation of alloys, or the combination of the base metals with the precious ones.

4. To obtain a single button of metal, which otherwise would be diffused in small globules.

The largest yields in reducing are obtained by first bringing the compound to be reduced, as well as the flux, to the finest possible powder and the most intimate mixture. Already from the earliest times the ingenuity of individuals has been taxed to produce mills or crushers for grinding the ores containing gold and silver to the finest possible powder, as it has been abundantly proved that this result is necessary to produce the largest yields of the precious metals. All residues, previous to fluxing, should be pounded to the finest powder, in a large iron or steel mortar. The proper selection, proportion and mixture of the fluxes is the whole secret which leads to the most successful reduction. The most important single substances used as flux are the following:

Borax or Biborate of Soda.

An excellent and almost universal flux, because it has the valuable property of forming fusible compounds with nearly all the bases. It is consequently very useful in purifying buttons of silver or gold, combining with and extracting the base metals, and rendering both the silver and gold perfectly pure. To prepare the commercial borax as a flux, it must be freed of its water of crystalization, which is best done by half filling a sand crucible with the borax and exposing it to heat until it is perfectly liquid, when it should be poured upon a large plate or stone, and, when cool, broken into fragments and preserved in well stoppered bottles.

Carbonate of Potash and Carbonate of Soda

possess great advantages over other single fluxes, as they are very efficient oxidizers and de-sulphurizers, and are very economical. Carbonate of potash is much preferable to soda, as it gives a very liquid and easily flowing glass, at a comparatively low temperature. On account of its great fusibility, it possesses the power of holding in suspension a large quantity of insoluble matter, as earth, charcoal, etc., etc. Carbonate of soda has an advantage over potash in not deliquescing or absorbing water from the atmosphere, and, as a mixture of the two, is not open to this objection; it combines the advantages of both. The best proportion is:

Carbonate of potash - - -	15 ounces.
Carbonate of soda - - -	10 "

Nitrate of Potash.

Nitrate of potash, or saltpeter, is used very extensively for refining and purifying gold and silver. It has a great tendency to oxidize, and consequently to remove the common metals. This gives us a very efficient method of rendering gold or silver perfectly pure; it is quite as excellent for refining as borax, and costs but one-half. Saltpeter is likewise very useful, added in small lumps, to burn out the carbon when reducing the paper ashes, which tends to make the flux "stiff."

Common Salt (Chloride of Sodium).

Common salt is of much use in silver reductions, as it checks, to a great extent, the tendency of the fused mass to boil over, and renders most fluxes much more easy flowing and liquid.

Black Flux.

Black flux is both a fusing and reducing agent. It consists of a very intimate mixture of carbonate of potash and charcoal. It is a very useful flux for chloride of silver and many other substances. To prepare it, mix intimately:

Cream of tartar - - - -	16 ounces.
Saltpeter - - - - -	8 "

Place the mixture upon an iron pan and ignite it with a red hot coal. When the combustion is completed, the black mass must be powdered and sifted while yet hot, and placed into a well stoppered bottle, as it rapidly absorbs moisture from the air.

Resin.

A most excellent flux for the chloride of silver is resin. It should be finely powdered and mixed intimately with the chloride of silver. It has been proved that before the chloride of silver attains a temperature sufficient to obtain a reduction, small portions of it are volatilized. By fusing with resin this difficulty is entirely overcome.

Molasses, Soap and Sugar.

These form excellent fluxes for the chloride of silver. Scrapings or raspings of Castile soap answer the purpose admirably.

We have given this rather lengthy list of fluxes, so that after a little experience the reader may be able to select for himself the substance, or combination of substances, most suited to his convenience, apparatus and facilities. We will now proceed to describe the method which we have adopted for reducing residues and know by experience to give good results.

FLUXING.

Having reduced the paper clippings and filters to ashes, the next step which should be taken is to sift them. In the paper wastes of photographic establishments, there is always a large amount of impurity, which renders the flux stiff and the results impure. The ashes should be rubbed to powder and thrown upon a fine flour sieve. The glass, nails, etc., may then be readily picked out. The residue which will not pass through the sieve, should be thrown into a mortar and powdered as long as any coarse grains remain behind. When the ashes are rich in silver, spangles of metallic silver are frequently left upon the sieve; these should be added to the ashes. Having concluded this operation, the following flux should be prepared:

Carbonate of potash - - - 16 ounces.
Carbonate of soda - - - 4 "

Mix the ashes intimately with their own weight of flux; fill a Hessian crucible about three-quarters full, with the resulting mass, and strew a thin layer of salt upon the surface. The crucible and contents are now ready for the fire.

Developer Washings.

When the developer drainings have been prepared by adding a solution of common salt, the black powder being thoroughly dried, should be mixed with its weight of the following flux:

Carbonate of potash - - -	10 ounces.
Nitrate of potash - - -	2 "

It should be placed into the crucible, precisely like the former. This flux flows easily, gives a fine glass and takes up all impurities.

Collodion Film Ashes.

The ashes of the collodion films consist chiefly of iodide and bromide of silver and some organic matter. The best flux for the reduction of these wastes is the following:

Carbonate of potash - - -	8 ounces.
Carbonate of soda - - -	1 ounce.

Add the flux in the proportion of five ounces of the ashes to four of the flux, and treat as the last.

Flux for Sulphide of Silver.

The sulphide of silver is the most difficult of the compounds of silver to reduce. When it has been previously roasted, as directed on page 49, the following flux, mixed in proportion of fourteen ounces of the sulphide to sixteen of the flux, answers excellently:

Flux for Roasted Sulphide of Silver.

Carbonate of potash - - -	15 ounces.
Carbonate of soda - - -	10 "

When it is desirable to do away with the first heating, the subjoined formula answers very well:

Carbonate of potash - - -	16 ounces.
Iron filings - - - -	2 "

Mixed in proportion as the last.

The last named flux requires a much higher heat than the former.

A flux which is superior to all others for the reduction of sulphide of silver, is the salt obtained by the evaporation of the waste cyanide fixing solution, as recommended on page 53. To every sixteen ounces of the sulphide of silver, twelve ounces of the cyanide salt should be added, and when thoroughly mixed together, a crucible is filled with the black mass. In this case the crucible should *not* be more than half full, as the mixture seathes very violently when exposed to a red heat. The result of the reduction is pure metallic silver and sulpho-cyanide of potash. This process should only be made use of when it is not desired to regain the cyanide of potash, as by these means a still larger economy is obtained.

The Toning Bath Precipitate—How to Flux it.

The precipitate obtained by the addition of sulphate of iron to old toning baths consists, as we have before stated, of metallic gold, mixed with a large amount of iron and the oxide of iron. When but a small amount of the precipitate is at hand, the wet method of treating it is much to be preferred to fusion. If, however, a considerable amount of this class of residue has accumulated, it should be mixed with saltpeter in the following proportions:

Gold precipitate - - -	16	ounces.
Saltpeter - - - -	12 to 14	"

The crucible into which the mixture is placed should likewise be filled but half full. On being exposed to a bright red heat, the gold is obtained in a nearly pure condition, the iron being oxidized and dissolved by the saltpeter.

Chloride of Silver.

The chloride of silver obtained by the precipitation of old baths, from the washings of prints, nitric acid plate cleaning solution and cyanide solution, may all, after being well washed, be mixed together previous to reduction. The chloride, having been exposed to a brisk heat, in order to expel any residuary moisture, should be mixed with half its weight of the following flux:

Carbonate of potash - -	16 ounces.
Powdered resin - - -	2 "

The crucible may be tightly packed three-quarters full, and a thin layer of salt strewn upon the top.

The residues having been fluxed and placed in the crucibles in the manner described, are now ready for reduction.

REDUCTION.

The furnace having been properly set, and the necessary appendages provided, as directed on page 38, a piece of brick, of about the size of the bottom of the crucible, should be placed in the middle of the grate, so that it does not obstruct much of the draft. The crucible is set firmly upon this brick, and shavings and thin kindling-wood filled up to its mouth and ignited. When the wood is thoroughly ablaze, charcoal, broken into the size of a large walnut, is thrown upon it. The crucible may now be covered, and the coke, or mixed fuel, filled right over its top.

When the fuel has burned away, it should be poked down around the sides of the crucible, and more added as quickly as it is required. If, when the contents of the crucible is liquid and thoroughly fused, any of the residue is left, it may be done up in thin paper, in the form of a cartridge, and added occasionally until the crucible is full; or the residue may be added with an iron spoon held by the tongs. Should, upon its addition the fused mass boil over, it may be quickly cooled down by stirring it with a cool poker or rod of iron.

When the last portion of the residue has been added, the fire should be raked, fresh fuel added, and the heat allowed to reach its highest intensity. After about thirty minutes' time, the reduction will be complete. To ascertain this to a certainty, lift off the cover of the crucible, and if the fused mass answers to the following tests the operation is completed:

1. The mass must be perfectly liquid and almost at a white heat.

2. When stirred with an iron wire it must *feel* liquid, and not as if full of sand or grit.

3. When a cool iron is plunged into the liquid mass and withdrawn, an even or smooth black glass or slag must be formed upon it, not one rough or sandy.

4. On dipping an iron wire to the bottom of the crucible, the fused silver can be *felt* by its resistance, and by a bubbling or rumbling feeling in the rod or wire.

If the fused mass does not answer these tests, th cover should be replaced upon the crucible and the heat again allowed to reach its maximum.

Where the contents of the crucible answer all these tests the pot may be removed from the fire and allowed to cool, or the contents may be poured into an iron mortar greased with a little lard or tallow. A crucible, when of a good quality, may be used twice, or even three times, with safety.

We have thus given the chief directions necessary, but there are of course still many little points and knacks which it would be impossible to describe, and which can only be gained by experience and practice. There are also many little precautions necessary in order to avoid loss of silver. Thus the coke or mixed fuel should be always plentifully supplied, and packed down around the crucible as quick as it becomes consumed, as if a current of cold air strikes the crucible while red hot, it is almost sure to crack it; then again, in poking the coals around the pot, it is very liable to crack, or when the residue contains much iron, it burns a hole right through inferior crucibles.

In adding the residue in wads, care is also necessary, as the fused mass, especially if at a high temperature, readily boils over, or loss may be occasioned by not adding the residue carefully or allowing the paper wad to catch fire before it reached the crucible. All these and many other little points are necessary to insure the largest yields, but they are readily learnt by a little practice. When the melted mass is not poured into a mold, but is allowed to cool in the pot, the crucible should be gently tapped with a piece of wood, so that any globules of silver retained or diffused in the glass or flux, may be separated or forced to the bottom. It must then be left undisturbed to cool. When perfectly cold it must be broken open with a hammer and the gold or silver will be found at the bottom, in the form of a lump or button.

If the reduction has been well performed the silver is obtained in one lump, and no globules of silver would be found diffused through the flux, while directly the contrary is the case if the operation has not been properly conducted. When paper ashes are reduced, the flux is very liable to become stiff or thick; it is then advantageous to throw upon the redhot mass a lump of saltpeter of the size of a filbert. By these means some of the impurities are burnt out.

REFINING, PURIFICATION AND GRANULATION.

Only the silver obtained from the chloride can be considered perfectly pure; that obtained from the sulphide ashes, or developer drainings, being always more or less contaminated with iron, or other impurities. If intended for the production of nitrate of silver, purity is indispensible, while if the metal is to be sold, it always brings a much higher price. The impure silver, as obtained from the residues, is placed in a melting-pot and brought to thorough fusion. Saltpeter, in small lumps, should then be thrown upon the redhot mass, in about the following proportion:

Metallic silver - - - -	8 ounces.
Saltpeter - - - - -	1 ounce.

When the violent deflagration ensuing has ceased,

the heat should be kept at its greatest intensity for fifteen or twenty minutes longer. The crucible is then removed from the fire, and its contents quickly and dexterously poured into an iron mold; or the pot may be allowed to cool, and finally broken open with a hammer. The button of silver thus obtained, after being thoroughly cleaned from all adhering flux (if intended for the manufacture of nitrate of silver), must be granulated. For this purpose place the pure buttons of silver, obtained as last described, into a clean crucible, without flux, and expose to an intense heat. When thoroughly melted, remove the pot from the fire, and slowly, but steadily, pour its contents into a pail of water, from the height of three or four feet. In pouring the silver into the water, the crucible should be gently swayed from side to side, so that the globules of the melted metal may be more thoroughly separated. The spangles of silver thus obtained should be carefully dried and kept in a clean stoppered bottle for future use. The process of granulation is only necessary when the silver is to be used for chemical purposes. It should always be *sold* in the form of a bar or button.

If the metallic silver has a bright and clean appearance, and is free from all adhering flux, it always fetches a higher price than the *purest* silver in a less attractive form. To obtain the silver as a bright button, the fused mass should be allowed to cool completely before any attempt is made to knock off the adhering flux. The button may be further cleaned and brightened by placing it, when cold, into the following solution:

Chemically pure hydrochloric acid	16 ounces.
Water - - - - - -	4 "

Or it may be boiled in a solution consisting of:

Bitartarate of potash (cream of tartar) - - - - - -	8 ounces.
Water - - - - - -	18 "

When the desired result has been obtained, the button should be wiped dry with a cloth, or it may be further improved by scouring with fine emery or sand mixed with a little water. What we have said of silver applies equally well to gold.

PRACTICAL SUGGESTIONS.

Old or cracked crucibles may be used with much success as stands or supports of the melting pot in the furnace. It is superior to a brick placed upon the grate, as, being very thin, they are very quickly heated through, and tend to equalize and distribute the heat around the crucible containing the residue. In removing the crucible from the fire, the hand should be protected by a pair of buckskin gloves, so that the strength of the grip may not be diminished by the intense heat. The operator should never omit, after having removed the pot from the fire, to thump the sides of the crucible, especially if the flux be stiff, as by these means any stray globules of silver are caused to run together. As a mold, a small iron crucible is the most convenient vessel. It should be previously heated, so that it cannot be touched by the hand, and lightly greased inside with a little lard or tallow. If an iron mortar is not at hand, a small vessel may be bent out of heavy sheet iron, which answers the purpose perfectly well.

A small stock of the fluxes should always be kept on hand, in well stoppered bottles, so that they can absorb no particle of moisture.

Considerable skill is requisite to produce well granulated silver. A very excellent method is to pour the melted metal through a *broom*, as by these means the particles of silver are separated and a finer division of the metal is insured.

We append a list of the most esteemed fluxes, from which the reader may select the one best adapted to his requirement.

FORMULAS FOR FLUXES, ETC., ETC.

For Ashes.

No. 1.

Carbonate of potash - - -	24 ounces.
Carbonate of soda - - -	6 "
Common salt - - - -	2 "

To be mixed with the ashes, weight for weight.

No. 2.

Carbonate of potash - - -	20 ounces.
Carbonate of soda - - -	18 "
Common salt - - - -	8 "

Fourteen ounces of the flux to be added to sixteen of the ashes. Requires a higher heat than the former.

No. 3.

Carbonate of potash - - -	15 ounces.
Silver ashes - - - -	16 "

Answers every requirement of a good flux.

No. 4.

Carbonate of potash - - -	8 ounces.
Nitrate of potash - - -	4 "

Renders the result pure by removing the iron, added weight for weight.

No. 5.

Carbonate of potash	- - -	8 ounces.
Carbonate of soda	- - -	8 "
Fused borax	- - - -	8 "

One pound of this flux is sufficient for every pound of ashes.

No. 6.

Caustic potash	- - - -	16 ounces.
Silver ashes	- - - -	16 "

A very excellent flux, if used *immediately;* but absorbs moisture and becomes liquid in a very short time.

For Developer Washings, Prepared by First Method.

No. 1.

Nitrate of potash	- - -	8 ounces.
Carbonate of potash	- - -	2 "

Add three-quarters of a pound to every sixteen ounces of dried precipitate.

No. 2.

Nitrate of potash	- - -	8 ounces.
Carbonate of potash	- - -	8 "
Carbonate of soda	- - -	4 "

In proportion same as last.

No. 3.

Nitrate of potash	- - -	10 ounces.
Dried precipitate	- - -	12 "

No. 4.

Borax	- - - - -	16 ounces.
Carbonate of potash	- - -	4 "

Added weight for weight. Inferior to the preceding formula.

No. 5.

Carbonate of soda	- - -	14 ounces.
Cyanide of potash	- - -	4 "

Very good, but expensive. Twelve ounces of flux to each pound of residue.

For Old Collodion Films, etc., etc.

No. 1.

Nitrate of potash	- - -	4 ounces.
Carbonate of potash	- - -	2 "

Half as much flux as ashes.

No. 2.

Caustic potash	- - - -	4 ounces.
Caustic soda	- - - -	2 "
Nitrate of potash	- - -	1 ounce.

In proportion same as last. Is a superb flux, but absorbs *moisture* almost instantaneosly when exposed to the air.

No. 3.

Carbonate of potash	- - -	4 ounces.
Borax	- - - - -	4 "

Added to the ashes in the proportion of twelve ounces of residue to ten ounces of flux.

No. 4.

Caustic potash	- - - -	8 ounces.
Nitrate of potash	- - -	8 "

Half a pound of flux to even pound of ashes.

No. 5.

Carbonate of potash	- - -	8 ounces.
Cyanide of potash	- - -	2 "

Mixed in proportion same as last. Flows easy and renders the result pure by oxidation.

For Chloride of Silver.

No. 1.

Carbonate of potash - - -	16 ounces.
Powdered resin - - - -	2 "

Add half the weight of the chloride

No. 2.

Carbonate of potash - - -	8 ounces.
Chloride of silver - - -	16 "

No. 3.

Chloride of silver - - -	8 ounces.
Resin - - - - -	12 "

No. 4.

Carbonate of potash - - -	8 ounces.
Carbonate of soda - - -	4 "

Proportion the same as the first.

No. 5.

Caustic potash - - - -	8 ounces.
Chloride of silver - - -	16 "

No. 6.

Chloride of silver - - -	4 ounces.
Black flux - - - -	3 "

No. 7.

Carbonate of potash - -	4 ounces.
Bitartarate of potash - -	2 "
Chloride of silver - - -	8 "

For Toning Bath Precipitate.

No. 1.

Nitrate of potash - - -	8 ounces.
Carbonate of soda - - -	4 "

Added to the residue weight for weight.

No. 2.

Nitrate of soda - - -	8 ounces.
Residue - - - - -	4 "

No. 3.

Nitrate of potash - - -	8 ounces.
Carbonate of potash - - -	4 "
Residue - - - - -	7 "

For Sulphide of Silver.

No. 1.

Carbonate of potash - -	8 ounces.
Carbonate of soda - - -	8 "
Slacked (well-dried) lime - -	4 "

This flux should be added, weight for weight, or even a considerable increase of this proportion may be made use of. The formula given above is very excellent, but the mixture requires a high heat.

No. 2.

Caustic potash - - - -	8 ounces.
Carbonate of soda - - -	4 "

Mixed with the residue weight for weight, flows easy at a low temperature.

No. 3.

Carbonate of potash - -	8 ounces.
Cyanide of potash - - -	2 "
Caustic soda - - - -	4 "

Very excellent, though rather expensive.

No. 4.

Carbonate of potash - - -	8 ounces.
Iron filings - - - -	1 ounce.

OR,

Carbonate of potash - - -	12 ounces.
Iron filings - - - -	2 "
Caustic soda - - - -	4 "

Proportion same as last.

No. 5.

Carbonate of potash - -	16 ounces.
Iron filings - - - -	2½ "
Saltpeter - - - - -	4 "

Very excellent.

No. 6.

The cyanide salt obtained by the evaporation of the cyanide fixing solution, mixed with four ounces of caustic potash to each pound of the salt.

This is one of the best known fluxes for the sulphide of silver. It flows easy and may be added to the *unroasted* sulphide.

Though the foregoing enumeration of fluxes may seem unnecessarily extended, still popular taste is so varied that where one suits, three may be rejected. We, for our part, see no necessity of making complicated mixtures of diverse substances, but prefer and make use of the plain carbonate of potash for all and every residue. The amount of potash added is varied according to the composition and nature of the wastes. For paper ashes, one and a half pounds potash to every pound of ashes may be mixed together; for chloride of silver, one-half its weight of potash, etc., etc.

CHAPTER VI.

REDUCTIONS BY THE WET OR GALVANIC PROCESS.

THOUGH it is true there are few residues which can be advantageously reduced by the wet method, yet the processes we are about to describe are so simple and economical, that the practical photographic operator and others who can devote little time and attention to this subject, will find these methods of reduction better adapted to their wants than the fire reduction. This is especially the case if the silver regained is to be used for chemical purposes, as by these means the metal operated upon is always obtained in a state of extremely fine mechanical division—a form peculiarly favorable to chemical action. Were it not that the wet processes are limited to certain compounds, they would entirely supersede the fine reductions for manufacturing purposes.

In this chapter we purpose to give the best methods for the reduction of residues without the use of fire. Reduction, strictly speaking, is the transmutation of a metallic compound to the metal itself; but under this head we have included the processes for regaining the silver in a useful form by merely chemical means, and without reducing them to the metallic state. We can, consequently, divide these processes into two classes: 1st. Those processes based entirely upon chemical action, and in which the metal contained in the residue is converted into a useful compound. 2d. Those dependent upon chemical action and electricity or galvanism, in which case the metal is obtained in a metallic or uncombined condition.

In the first case, the operation consists simply in pouring upon, or digesting, the gold and silver residues in some solution which has the power of combining with the precious metal itself, or the substance with which it is combined, thus altering and modifying its chemical properties. In the second case, namely, by the action of electricity, the metal, as before stated, is obtained in the metallic condition, while the solvent action of the chemical added is confined to the substance used to generate this electricity.

TREATMENT OF WASTES BY CHEMICAL ACTION.

Paper Clippings, Filters, Filtering Cotton, etc., etc.

For successfully treating this class of wastes by the wet method, a thorough combustion of the paper is of primary importance. For this purpose proceed as directed on page 46. When the ashes have cooled they must be carefully sifted through a fine flour sieve, and all impurities removed, while the globules and spangles of metallic silver remaining in the sieve are added to the sifted mass. Place the gray powder thus obtained into a spacious porcelain evaporating dish, and for each pound add twenty-four ounces of the following mixture:

Pure concentrated nitric acid	-	16 ounces.
Distilled water - - - -	-	6 "

and, after stirring thoroughly with a glass rod, set the

dish and contents in a vessel of hot water, or upon a moderately warm stove. After the violent ebullition and all red vapors have ceased, bring the liquid to the boiling point and let it simmer about fifteen minutes; then dilute the dark muddy mass with an equal bulk of distilled water, throw the whole contents of the dish upon a large paper filter, and carefully collect the clear liquid trickling through. If the filtrate still looks a little murky or unclear the filtration may be repeated. Wash the residue remaining with a little distilled water, and add this filtrate to the other. The solution thus obtained consists of moderately pure nitrate of silver. It should b placed in a porcelain evaporating dish and boiled to dryness on a sand bath. When all of the water has evaporated the heat should be still continued, until the mass of white crystals (under constant stirring) begin to assume a dirty brown color, and have fused to a dirty colored pasty mass. When this stage is arrived at, quickly remove the dish from the fire and allow it to cool perfectly; then add enough distilled water to effect a complete solution, or until by means of a hydrometer, you have brought it to the strength desired for your bath, and filter the murky liquid until it is perfectly clear. By these means a superb and perfectly neutral silver bath may be obtained. The black mass remaining in the filter still contains some silver, and may be dried and reduced or sold at the option of the operator. This process has been described in "Humphrey's Journal of Photography" by a gentleman who forgot to credit its origin to the author. If we have urged the necessity of keeping the residues clean for fire reductions, we must urge it in a tenfold degree if they are to be reduced by the wet method. If the paper is carefully kept and well burnt, the resulting silver is of a very fair purity; while if it contains nails, pins, or similar substances, the nitrate will of course contain a variable amount of the nitrates of iron, copper, etc., etc., substances highly detrimental to a good bath for sensitizing photographic plates. But with a little careful management the paper can as well be kept perfectly clean as mixed with foreign substances.

Treatment of Developer Drainings.

Place the precipitate obtained, as directed on page 48, with common salt (that obtained with sulphide of potash will not answer) upon an iron shovel; set it upon a lively coal fire, and expose it to a red heat for about fifteen minutes. When the now brick red powder has cooled, place it in a capacious evaporating dish, and for every ounce of the powder add an ounce of pure hydrochloric acid, previously diluted with half an ounce of water, and, after stirring the mixture, place it upon a stove and allow it to gain the boiling temperature. When all violent effervescence has ceased, add a little more acid, and if no further action takes place the operation is complete. Dilute the dark liquid copiously with water and allow the black precipitate to settle. Pour or decant off the yellowish green solution (which consists of the chloride of iron, and is of no value whatever); throw the black precipitate upon a filter, and, after washing copiously both with hot and cold water, allow it to dry; it consists of nearly pure metallic silver. This may be made into the nitrate, as directed in another chapter. The heating of the powder may be dispensed with, but more acid is then required for the solution of the iron and the results are never so reliable. The hydrochloric acid used in this process need not be chemically pure, and, if very concentrated, should be mixed with an equal bulk of water; but if only moderately strong, not more than half its bulk of water need be added. This process is not always reliable, and we would thereupon prefer the fire reduction. If, for instance, the acid is too strong, some of the metallic silver is changed to the chloride; if too weak, some iron remains behind.

Chloride of Silver—How to Change it into Nitrate.

To change the chloride into the nitrate of silver is a very simple and economical process, to perform which proceed as follows:

Place recently precipitated chloride of silver into a large glass flask, and to each ounce add two ounces of a concentrated solution of caustic potash; set the flask upon a sand bath until the solution begins to boil. A marked change will now be observed, the white chloride of silver will rapidly change to a dark brown, and when no further darkening takes place, remove the flask and dilute its contents with plenty of water. Collect, wash and dry the dark brown powder and preserve in well stoppered bottles. It consists of chemically pure oxide of silver, and may be used either for the manufacture of nitrate of silver, or, if only small quantities have been obtained, for neutralizing and restoring acid silver baths.

Another method which answers tolerably well, is to

pour upon the chloride a concentrated hot solution of carbonate of potash and boiling the mixture as before. The resulting brownish powder should be treated precisely like the former. It consists in this case of carbonate of silver. In both these processes it is absolutely indispensable, that the chloride of silver be freshly precipitated, or at least not allowed to dry; this may be easily accomplished by immersing the precipitate in water, and keeping it submerged until enough has accumulated. (See page 55.) Care should also be taken to keep the chloride of silver perfectly clean. The nitrate of silver solutions to be precipitated should be previously filtered perfectly clear, the bottle used as receptacle be properly cleaned and always kept corked in a dark corner of the sensitizing-room. When these precautions are observed the results, especially of the first process, are *chemically pure*, and the resulting nitrate of silver is consequently of a very superior quality.

To Change Iodide or Bromide of Silver into the Nitrate.

A process very similar to the former. Place the recently precipitated, and of course still wet mass, into a flask, and upon each ounce pour two ounces of a hot and very concentrated solution of caustic potash. Place the flask and contents upon a sand bath, and allow the liquid to boil for about half an hour, adding a little water if the solution boils away too rapidly. At the end of this time dilute the contents of the flask with abundance of water; decant the liquid and filter off, and wash the precipitate thoroughly.

The result is also oxide of silver. We consider this

process rather uncertain, but have often obtained good results with it. The iodide and bromide of silver is of such rare occurrence, however, as a residue, that the process is of no practical value.

To Change the Print Washings into Nitrate of Silver.

The print washings consist when precipitated, as we have seen, of pretty pure chloride of silver, and may consequently be treated in precisely the same manner; more care is however necessary to keep washings clean as slips of paper, chips of wood, etc., etc., all tend to make the results impure. The barrel or jar containing this residue, should be kept well covered over from dust, etc.

How to Treat the Toning Bath Residue.

The precipitate of the toning bath obtained, as directed on page 54, by the addition of a solution of sulphate of iron, can be much more advantageously reduced by the wet method than by fire, especially as there can be but little collected even in the largest galleries.

When the ebony black powder has been dried, place it upon a shovel or other iron vessel, and expose it to a red heat on a coal fire for at least ten minutes. The contents of the shovel will now be found to have changed from jet black to dark red, owing to the liberation of sesquioxide of iron. Place this red powder in an evaporating dish, or a large flask, and pour upon it the following mixture (three or four ounces of acid to one of precipitate):

Strong hydrochloric acid - -	8 ounces.
Water - - - - - -	6 "

Place the dish upon a stove and heat it until all effervescence has ceased and the liquid begins to boil; then dilute with plenty of water, and filter, wash and dry the black powder, which consists of pure metallic gold in a state of minute division. The gold must be washed very thoroughly while upon the filter (both with hot and cold water), to ensure the removal of the chloride of iron formed. The acid used need not be chemically pure, but care should be taken that it contains no *free chlorine*, as this would dissolve the gold and cause a considerable loss. To test acid as to its containing free chlorine, proceed as follows:

Place a few drops of the acid upon a watch-glass, and lay in these a minute piece of real gold leaf. If the gold leaf disappears, the acid should be rejected; while if it remains in the liquid without being dissolved, the acid may be used. Or the following mixture may be made use of, though it is much inferior to the former:

Sulphuric acid - - - -	8 ounces.
Water - - - - -	16 "

Three or four ounces of acid should be added to each ounce of precipitate, and the mixture treated precisely like the former.

Old Solutions of Nitrate of Silver, etc., etc.

When a bath is rejected, and is no longer reliable in its action, it should be placed in a large flask or other suitable vessel, and the following solution added as long as it causes a precipitate:

Carbonate of soda - - -	8 ounces.
Water - - - - -	12 "

OR,

Caustic potash - - - -	4 ounces.
Water - - - - -	12 "

Then add water, wash the precipitate thoroughly and abundantly with warm water, and finally dry and preserve it in well stoppered bottles. The first precipitate consists of *carbonate* of silver, the second of oxide of silver.

A solution of lime is also a very excellent precipitating solution for silver. Place some lumps of lime in a wide mouthed bottle and add considerable water; allow the mixture to remain at rest for several days, and then drain off the clear liquid that has gathered on the top. This solution possesses many advantages over potash or soda.

REDUCTION OF SILVER COMPOUNDS TO THE METALLIC STATE BY GALVANIC ACTION.

When soluble metallic salts are treated with chemicals that generate electricity, or are exposed to a current from a battery, they are invariably decomposed and changed to the metallic condition. This decomposing and reducing action of galvanic or voltaic electricity, affords us a convenient and efficient method of utilizing or regaining the silver from waste silver solutions; but its application is also limited to a few compounds, and here, as in the foregoing processes, sulphide of silver is the most obstinate, and the chloride the most tractable.

There are several methods for successfully treating the chloride of silver by galvanic action, and we will proceed to describe the simplest of these.

1. Place recently precipitated or still wet chloride of silver in a porcelain evaporating dish, in a cup, flask or any other convenient vessel, and mix with it half its weight of small pieces of zinc. Then add enough of the following solution, to cause a very lively effervescence, and let the mixture stand over for a day or so:

Sulphuric acid (oil of vitriol)	-	8 ounces.
Water - - - - -	-	4 "

Should any pieces of zinc be left after twenty-four hours' standing, more acid should be added *until they entirely disappear*. Then dilute the now black mass with water, slightly acidulated with sulphuric acid, and, after agitating the mixture for a short time, let the solution remain at rest until the whole of the black powder has settled to the bottom. Now decant the water, throw the black muddy mass upon a filter and wash copiously with hot water. Finally, dry the contents of the filter and the result is *pure metallic silver* in a state of minute division. It should be carefully preserved in well stoppered glass bottles from dust and other foreign substances. When this process is conducted properly the result is chemically pure. Great care is however necessary to continue the washing long enough, as otherwise the metallic silver obtained is contaminated with zinc, and is entirely unfit for photographic purposes.

2. Into some freshly precipitated chloride of silver, hang strips of copper or throw some copper coin, and acidulate the mixture with a few drops of oil of vitriol. After twenty-four hours' standing the copper will be incrusted with metallic silver; but the copper should

remain in the mass until all of the chloride is transmuted to the metallic state, which can be readily ascertained. Then like the former the mass should be diluted with plenty of water, the gray metallic powder allowed to settle to the bottom, and the incrustation removed from the copper strips, the whole mass thrown upon a filter, washed with the acidulated water, and finally with pure hot water and dried. The result is, as in the former case, pure metallic silver.

In both the foregoing processes the chloride should be mixed to about the consistency of *cream* with water.

3. Instead of using strips or small pieces of zinc, substitute fragments of iron. Proceed precisely as in formula first, and the metallic silver obtained will be very pure.

4. Place the chloride of silver to be reduced in a convenient vessel, and throw in it a strip of copper connected with the zinc end of a battery. In a day or so the whole of the chloride of silver is reduced to the metallic state, in the form of a beautiful incrustation on the copper.

5. The chloride of silver may easily be reduced to pure silver without the aid of fire. Keep the salt under water in the dark-room until you have enough to operate upon, and time to devote to the operation. The following is the mode of reduction:

Take a bar of clean zinc as heavy as the quantity of chloride to be reduced, and solder to one end of it a silver wire; then cover the zinc completely with fine gauze or muslin and dip it in clean water. Now immerse the zinc, so covered, in the moist chloride of silver, and bend over the other end of the silver wire, so as to come in contact with the chloride of silver at a short distance from the remote end of the zinc. The operation is best conducted in the dark-room. The moment the connection is made with the silver wire and the chloride, an electric current sets in and decomposes the chloride of silver into pure silver, which manifests itself first at the loose end of the silver wire. The chlorine which is set free hastens through the muslin and combines with the zinc, forming chloride of zinc, a very soluble salt which remains in solution. The operation may continue until all the white chloride has changed color and become silver gray. The bar of zinc is now taken out and washed to remove any adhering silver; it is much lighter than it was before the operation. Dilute sulphuric acid is added to the silver powder in order to dissolve any particles of zinc; after settling a number of hours the supernatant liquid is poured away, and the residue is well washed in several changes of water. The residue is pure silver containing still, probably, some undecomposed chloride of silver which is no injury to it.—*Photographer's Guide.*

This is a rather unnecessarily complicated arrangement, and the results are not as good or certain as with anyone of the foregoing ones. We give this description not as having our recommendation, but, on the contrary, we do so only to have the chapter as complete as possible.

Sulphate of Silver.

When old solutions, etc., etc., have been precipitated in the form of a sulphate, the still wet powder may be readily reduced to the metallic condition, for which proceed as follows:

1. Place the recently precipitated white powder in a bowl or evaporating dish, and upon each ounce pour two ounces of the following solution:

Sulphate of iron - - -	4 ounces.
Water - - - - -	12 "
Oil of vitroil - - - -	4 drops.

After thoroughly mixing the mass place it in a warm place and allow it to remain there for twenty-four hours. Then dilute the mass copiously with warm water, throw the result on a filter, and wash thoroughly first with hot water acidulated with sulphuric acid, and then with hot water alone.

The result is pure metallic silver, which is usually found in the mass in the form of little thin metallic tablets.

2. Connect the sulphate with the zinc end of the battery by means of a strip of copper, and let the action proceed until the whole of the sulphuric is changed to the metal.

Old Nitrate of Silver Baths, Waste Nitrate Solutions, etc.

To reduce old solutions, etc., of nitrate of silver to the metallic state, it is not necessary that they be previously precipitated, but they may be treated as follows:

1. Hang a broad and long strip of copper in the nitrate solution, and allow the silver to deposit on it for several days. Should the strip of sheet copper be consumed, hang in another one and allow the whole solution to remain at rest as long as any deposition

takes place. When this action has ceased remove the remaining copper from the solution, scrape off the adhering silver and wash the resulting powder as the last.

2. Repeat the above process connecting the copper strip with the zinc end of the battery, and the process will be finished almost in as many *hours* as it takes *days* in the former.

3. Repeat the first process, substituting zinc for copper. The result is not so good.

Neither the iodide, the bromide or the sulphide of silver can be treated by the galvanic current.

There is yet another way in which chloride of silver may be very economically treated in the wet way, and in which there is no contamination of the silver by another metal to be feared, it is as follows:

Place in a bowl or flask the recently precipitated chloride, and pour upon each ounce *two* to *three* ounces of the following solution:

Caustic potash - - - - -	8	ounces.
Water - - - - - -	12	"
Loaf sugar - - - - -	3	"

Boil the chloride in this liquid for about half an hour, or until the white chloride has changed to a dark brown or even black color; then dilute copiously with water, throw the black powder upon a filter, and wash thoroughly with water. Finally, dry and preserve in well stoppered bottles for future use.

The potash changes the chloride into the oxide of silver, and this, in time, is again transmuted by the reducing action of sugar into metallic silver.

In this case no electricity is excited, but the reducing action is, nevertheless, complete.

The Apparatus.

The apparatus required for the operations mentioned in the preceding pages is very simple, and of a precisely similar character as that used for treating the liquid residues preparatory to smelting. (See page 56.)

When the chloride of silver is to be reduced by the action of metallic zinc, a beaker or precipitating glass is a very convenient vessel. The zinc should be in the form of small tablets or plates, so as to expose a a large surface to the action of the acid.

The most convenient way is to use the granulated metal. For this purpose the zinc is placed in an iron spoon or ladle, and brought to thorough fusion, then poured into a pail of water from the height of about four feet. When copper is made use of it should always be in the form of a thin strip or tablet. The quickest and best way to reduce silver is, however, as already shown, by means of a galvanic battery, which

Fig. 10.

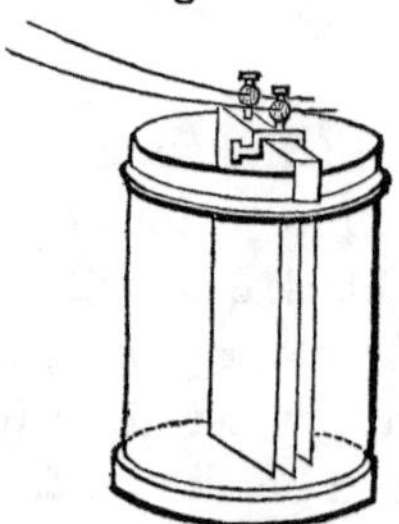

for this purpose may be of very simple construction. The one best adapted for reductions is that known as

Smee's, represented by Figure 10. It consists of two plates of zinc clamped to a piece of wood, and connected together by means of a strip of brass ending in a screw cup for the reception of the conducting wire.

Between the zinc plates (which should be about half an inch apart) is suspended a thin sheet of platinized silver, which is also connected with a screw cup, and forms the second pole of the battery.

This combination is hung in a proper glass vessel, which, when the battery is to be set in action, is nearly filled with the following mixture:

Sulphuric acid - - - -	2 ounces.
Water - - - - -	16 "

The chloride of silver of about the consistency of cream is poured into a convenient vessel, and in it is suspended two strips of copper, but *not touching* each other.

The chloride will now be seen to change rapidly in color, and a deposit of pure silver to collect on the copper attached to the zinc end of the battery, while the other strip is dissolved correspondingly fast.

This form of battery is the quickest, but one quite sufficient, for moderate quantities of wastes may be put together as follows:

Provide two pieces of zinc and two pieces of copper about 4 x 6 inches in size. Have one piece of copper connected with one piece of zinc by means of a thin strip or wire of copper soldered to each, and the other two pieces may have two separate wires about two feet in length attached to them. When this arrangement is complete, place the two pair of metal combinations in two separate glass or porcelain vessels, as represented by Figure 11. To set this little battery in op-

Fig. 11.

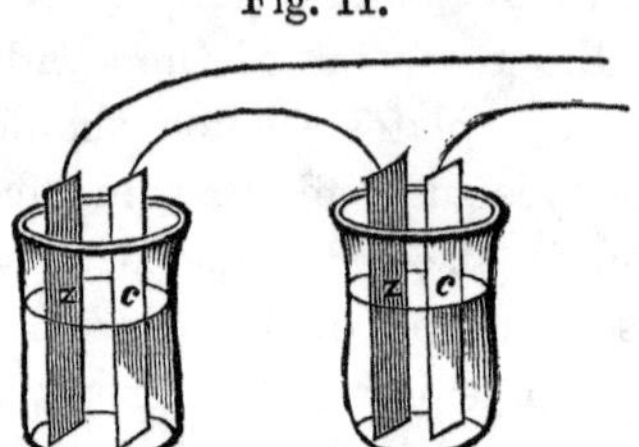

eration the cups are filled with strong salt water or dilute vinegar as follows:

Strong vinegar - - - -	4 ounces.
Water - - - - -	4 "

Connect the wires to two pieces of copper, as in the last, and let them remain in the chloride for about twenty-four hours, changing the salt water *once* during this time. A complete reduction is effected in from one to two days, while with a Smee's battery it is accomplished in a tenth of this time.

Great care must always be taken to wash the reduced mass with water, acidulated with sulphuric acid.

Stronger acid may be used, if a plug of asbestos is placed into the neck of the funnel as a filter.

CHAPTER VII.

QUANTITATIVE ESTIMATION OF THE PRECIOUS METALS CONTAINED IN WASTES AND ASHES.

The large extent to which fraud is committed in the reduction of gold and silver residues—aided and rendered easy by the photographer's lack of knowledge of the first principles of chemistry—has made it absolutely necessary that some mode of estimating the amount of the precious metal contained in his residue be made known to him. Though it is an easy thing for a chemist in his laboratory to test in a few moments, at least approximately correct, the amount of precious metal in a certain substance, still for an unexperienced person it is anything but an easy operation. When a simple substance, like the chloride, iodide, bromide, sulphide, or nitrate of silver is under consideration, the amount of *metallic silver* to be derived from anyone of these substances is readily determined, and is but a simple arithmetical problem. The chloride of silver has the formula of $AgCl$, and consists of an atom of silver combined with one of the gas chlorine; the bromide $AgBr$, one atom of silver and one of bromine; the iodide AgI, one of silver and one of iodine; the sulphide AgS_2, one of silver and two of sulphur; and finally, the nitrate $AgNO_5$, one atom of silver combined with one of nitric acid (NO_5), or otherwise, one atom of silver combined with one of nitrogen and five of oxygen.

To estimate then, for instance, the amount of *metallic* silver contained in a given quantity of the *chloride*, we proceed as follows:

The *atomic weight** of metallic silver is 108 and that of chlorine 35, then by adding 108 and 35 together, we obtain 143 as the atomic weight of *chloride of silver.* Finally, if we divide 108, the atomic weight of *metallic* silver, by 143, the atomic weight of the *chloride,* we have remaining in the quotient the decimal ·75524, which equals (allowing for loss in reduction) $\frac{75}{100}$ or $\frac{3}{4}$. Every sixteen ounces of chloride of silver are consequently equivalent to twelve ounces of metallic silver, or seventy-five per cent. of the chloride consists of metallic silver. Where, as in the sulphide of silver, there are two equivalents of the non-metallic body, the arithmetical process is the same, but the atomic weight of the non-metallic body must be doubled, as it has two equivalents. (AgS_2.) Where, as in nitrate of silver, there are two haloids, or non-metallic substances ($AgNO_5$), their atomic weights are added together, while the atomic weight of the last one, oxygen, is multiplied by five.

With the knowledge of the atomic weight of the component parts of a body, it is easy for us to calculate the amount of metal which they contain, and enables us to make the following table:

* If in a compound of one grain of hydrogen with eight of oxygen there be an equal number of atoms of each, an atom of the latter will be eight times as heavy as an atom of the former. In this way we know the relative weights of the atoms of all substances, whose equivalents are known.

TABLE *of the Amount of Metallic Silver contained in a Compound of Silver.*

Name of Compound.	Symbol.	Atomic Weight.	Theoretical Amount of Silver.	Actual Yield, shown by Weight and Percentage.
Chloride of silver....	$AgCl$	143	·75524	Three-quarters, or nearly seventy-five per cent.
Sulphide of silver....	AgS^2	140	·77142	Three-quarters, or seventy-five per cent.
Bromide of silver....	$AgBr$	186	·57746	One-half, or fifty per cent., generally a little more.
Iodide of silver......	AgI	234	·45524	Three-sevenths, or forty-three per cent.
Oxide of silver......	AgO	116	·93102	Nine-tenths, or ninety per cent., generally more.
Carbonate of silver..	$AgCO^2$	136	·79411	Three-quarters, or seventy-five per cent., generally a little more.
Nitrate of silver.....	$AgNO^5$	162	·63529	Two-thirds (not quite) sixty per cent. (full.)

This table, as will be immediately seen, applies merely to the pure and simple compounds of silver, and such residues as paper ashes, collodion films, developer washings, etc., etc., cannot be estimated in this manner, as they consist of several diverse substances. To find correctly the amount of metallic silver contained in ashes or other wastes not comprised in the above table, it is necessary to analyze a small portion and thus to test the precise amount by actual experiment. This may be done in several different ways, but the simplest is to mix a drachm or so of the powder under consideration with a little borax, say *weight* for *weight*, and to expose the mixture in a very little crucible (quarter ounce) to a brisk heat in an ordinary kitchen fire. The little pot should be well embedded in the glowing coals, and the reduction will be complete in from fifteen to thirty minutes. The little crucible should, when the desired result is attained, be removed from the fire, allowed to cool, and finally broken open with a hammer, when the metallic silver will be found at the bottom in the form of a little button or lump, which varies in size from the bead of a pin to a small pea. On accurately weighing the little button it is easy to calculate the amount of metallic silver to be derived from the amount of ashes the operator may have in his possession. Supposing the yield from one drachm is 35 grains, by multiplying by 8 we obtain 280 grains as the yield from one ounce of ashes, and by again multiplying by 16 we obtain 4,480 grains, and by dividing this sum by 360 (the amount of grains in one ounce) we obtain 12 ounces and 108 grains as the yield from one pound of ashes.

By these means the operator may readily ascertain, in a few moments, the precise amount of the precious metal contained in any one of his residues, and he may consequently readily defend himself against fraud or imposition on the part of the refiner.

But the process just given is open to several objections. In some galleries, for instance, there is no stove at hand; again, the manipulations are the same as in reducing large quantities, and there is almost as much trouble incurred in thus testing a drachm of the residue as in reducing a whole pound of it. These objections are, however, readily overcome by the use of a blowpipe.

"The blowpipe is a small and convenient instrument by which a blast of air may be forced through a lamp, gas-flame, or even a candle-flame, so as to intensify the heat of the latter to such an extent as to render it a substitute for the furnace in very minute and delicate operations.

"A simple form of the blowpipe, and that originally adopted for soldering, etc., by metal workers, is represented by Figure 12. It consists of a tapering tube of brass, curved nearly at a right angle a short distance from the smaller end. The hole terminates in a

Fig. 12.

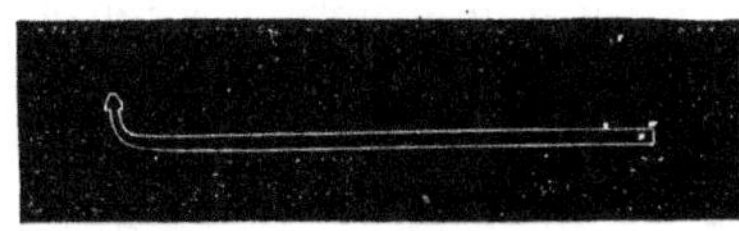

fine perforation. A steady stream of air is forced through the pipe by the action of the muscles of the cheeks, and directed against the flame of a candle or lamp."

There is considerable practice necessary to use the blowpipe successfully; but, when once learned, it is a pleasant and useful acquirement. To make the manner of using the mouth blowpipe, and acquiring the knack of *breathing* and *blowing at the same time* plain to the reader, we make an extract upon this subject from "Bowman's Practical Chemistry:"

"Before proceeding to any blowpipe experiments, it is necessary to acquire the knack of keeping up a constant and unremitting blast of air from the mouth, as without this it is impossible to raise the heat to a sufficient degree of intensity. The habit is readily acquired, and when once attained, the mouth and lungs will be found to do their work almost mechanically, without any sustained effort on the part of the operator.

"The learner may first observe that on closing the lips he can still without difficulty breathe through the nostrils: let him now distend the cheeks with air from the lungs, and he will find that on closing the communication between the mouth and throat he can breathe through the nostrils for a length of time, still keeping the cheeks distended. He may next introduce the mouth-piece of the blowpipe between his lips, and having puffed out his cheeks with air from the lungs, and again closed the communication between the mouth and throat, let him breathe freely through the nostrils, at the same time allowing the distended cheeks to force a current of air through the blowpipe. When the stock of air in the mouth is nearly exhausted, a fresh supply is sent up from the lungs, when the cheeks, again distended, will by their elasticity keep up a current of air through the blowpipe, while the operator breathes through the nostrils as before.

"The cheeks thus play the part of an elastic bag, with a valve opening inward, which, if connected with the blowpipe, and distended with air, would force air through it as long as the tension of its stretched sides exerted sufficient pressure."

Having now, we will suppose, acquired the knack of using the blowpipe, it is an easy matter to estimate quantitatively every existing gold or silver residue. Weigh out very accurately about fifteen grains of the residue, and add to it fifteen grains of a mixture of equal parts of carbonate of soda and cyanide of potash. Then obtain a piece of pine or willow charcoal and scoop a cavity in it, just large enough to contain the above mixture; then support the blowpipe on the edge of the flame of a lamp or candle, and blow a steady fine flame upon the mixture. (Figure 13.) The

Fig. 13.

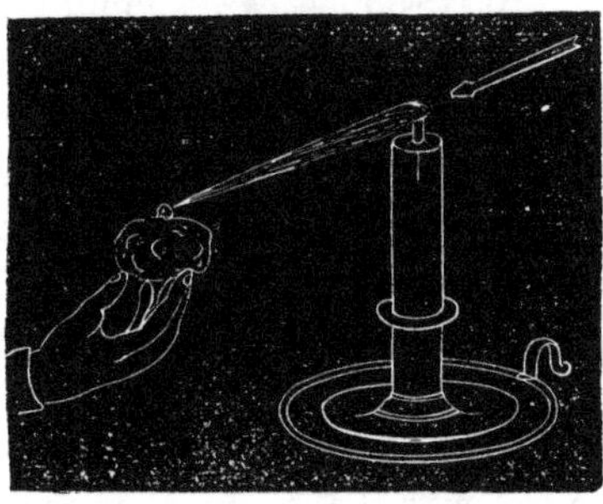

residue soon begins to glow, to melt, and bubble, and finally the globule of silver will be seen dancing about in the glowing cavity. When the reduction has been completed, allow the charcoal and contents to cool, remove the globule of silver by means of a little pair of nippers, and wash thoroughly from any adhering flux. By weighing the result on a delicate balance, the precise amount of metallic silver, contained in the residue on hand, may be estimated. No matter what the residue may consist of the flux in this operation should be always the same—equal parts of carbonate of soda and cyanide of potash. If the flame be skillfully managed the residue should be complete in about five minutes.

The great beauty and simplicity of the blowpipe process must be apparent to every reader. The trouble incurred is so small and the result is so accurate, that this mode of estimation should be made use of in every case where the residues are sent to smelters or refiners for reduction. By this process the photographer and silver worker may accurately estimate the value of waste sent to the refiner.

In estimating the residues, care should, of course, be taken to obtain a fair average of the sample under consideration. For this purpose the ashes or other wastes should be carefully sifted through a fine flour sieve, and all glass and other foreign matter removed, while any lumps of chloride of silver contained therein, are ground to powder and mixed evenly throughout the sifted mass.

There is no way of estimating, at least very accurately, by the wet way the amount of metallic silver contained in a residue.

When *arithmetically* calculating the amount of silver

contained in one of its compounds, we must, however, bear in mind that the *actual* yield is never so large as the *calculated* one, for, as a general rule, the residues are not pure, but, on the contrary, more or less contaminated with foreign impurities. Thus, for instance, the precipitate of the print washings, generally called *chloride of silver*, and which would consequently be *calculated* to contain $\frac{75}{100}$ or $\frac{3}{4}$ its weight of metallic silver, generally yields *actually* only fifty to sixty-five per cent. (according to the way it has been saved), on account of the large amount of albumen it contains.

Then again, when estimating about *fifteen grains*, as above directed, the reduction is necessarily much more complete than when pounds of the residue are reduced at once, and a small percentage should be allowed for loss in this way. In all cases where any doubt exists as to the *purity* of a compound (for only such can be arithmetically estimated), a small portion of the substance under consideration should be reduced with the blowpipe, as before directed, as then only a certain and correct estimation can be made. The impure chloride of silver obtained from print washings, can be very correctly determined without heat as follows:

Place one drachm (thirty grains) of the well dried print washing's precipitate in a small evaporating dish or other convenient vessel, and pour upon it one ounce of a supersaturated solution of caustic potash, and let it heat for a few minutes on a stove. When the chloride has changed to a dark brown or black, dilute the mass with eight ounces of water and let the precipitate settle; wash a few times in this manner, then throw upon a filter (better a tuft of cotton in the neck of a funnel), and let the liquid run through. Collect the brown powder carefully, dry it thoroughly, and if much remains upon the cotton tuft, burn this and add the ashes to the rest of the powder. When it is completely dry throw the powder on a delicate balance, and referring to our table for "Oxide of Silver" the amount of metallic silver can be readily determined. By the above process all of the albumen is removed by means of the potash, and the estimation is as correct as by fusion with the blowpipe.

To estimate the amount of silver contained in ashes the mode of procedure is as follows: Place half an ounce of the ashes (which for this purpose should be thoroughly burnt) in a small evaporating dish or a beaker glass, and pour upon these about two ounces of strong nitric acid and one ounce of water. Set the mixture upon the stove and let it boil rapidly nearly to dryness; then dilute the mass with four or five ounces of water, and filter through a plug of cotton placed in the neck of a funnel. The clear filtrate is placed in a clean evaporating dish and boiled rapidly to dryness.

The crystalline mass of nitrate of silver should be carefully dried and estimated according to "Nitrate of Silver" on our table; or it may be reduced very easily on charcoal, by a few puffs from the blowpipe, and the resulting metal carefully weighed; one can estimate approximately correct the amount of silver the ashes will yield. It will be seen by the foregoing processes that the estimation of *residues* by the wet way is nothing more than the reduction of a small portion of the same, and the blowpipe is merely a miniature furnace, in which all the operations of reducing are performed on a small scale. Many other ways of *estimating* residues by the wet way will be suggested to the reader by the perusal of Chapter VI., as, for instance, the estimation of developer residues.

PRACTICAL SUGGESTIONS.

The main thing necessary to obtain a correct estimate of the amount of silver contained in a substance is a delicate balance or scale, which should turn at least the one-twenty-fifth of a grain. A balance very suitable for weighing small quantities of silver, or the iodides and bromides in the preparation of collodion for photographic purposes, may be readily constructed with a little care and ingenuity as follows:

Have two little scale pans, two inches in diameter, punched out for you by the tinsmith, and put three holes in each of equal distances apart. A silk thread is tied into each one of these holes, and they are then knotted together, and a silken thread, three inches or so in length, connected to these. A thin straight piece of wire should now be provided, and one of the pans connected to each end of this wire. In precisely the middle of the wire should be secured a strip of wood, about the length and size of a match, and to this two pieces of silk thread, about three inches in length, which for convenience' sake may end in a brass ring. To illustrate this more clearly we will describe the annexed woodcut.

A A are the scale pans connected with the three silk strings B, which are connected to a hook on the wire C, which (*precisely*) in the middle is fastened the strip of wood D, from each end of which passes a silk thread E, which, in turn, connects with the single thread F. When the middle of the beam is correctly

Fig. 14.

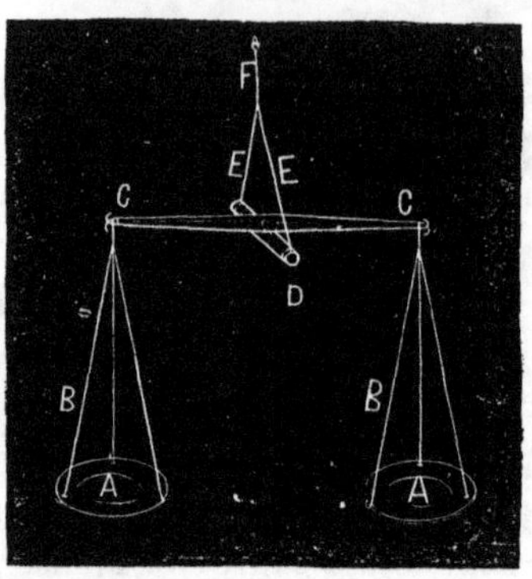

balanced, and the whole thing properly put together, this balance is of wonderful delicacy, and the pans will turn readily with the one-fiftieth part of a grain; and we once even constructed one on the same principle that turned with less than the one hundreth part of a grain. A balance of like delicacy could not be purchased for less than $50.

The balance should be so suspended that the pans are within a quarter of an inch from the table or stand. A convenient way to suspend the balance is to make a little stand from a pine board, and driving a wire with a hook at its end into it, as represented by the accompanying woodcut.

When not in use the balance should be covered from dust or dirt by a glass globe, or by a box of glass made by uniting panes of glass of the proper size by means of paper or cloth.

Fig. 15.

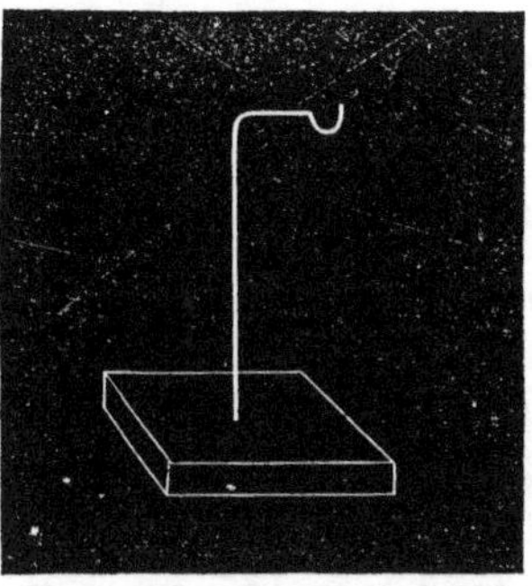

A very good blowpipe may be made in emergencies by accurately fitting the bowl of an ordinary tobacco pipe with a tight cork, and fitting into this a glass jet or an inch of the stem of a pipe having a *very minute bore.*

Fig. 16.

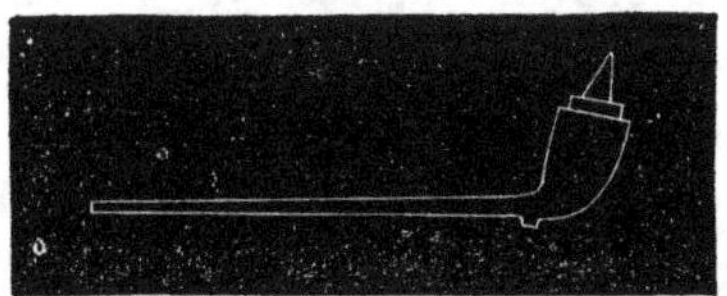

A glass jet is formed by drawing out a glass tube to a very fine point, and holding it into the flame until it thickens at the end. Finally, file a small hole into the top with an ordinary file. When residues are estimated a test-tube (a thin glass tube closed at one end) may be used instead of the more cumbersome evaporating dish.

CHAPTER VIII.

USEFUL HINTS AND FORMULAS.

We would in every case advise the practical photographic operator to manufacture nitrate of silver from the result of his reductions, in preference to selling it in the metallic form to a refiner, as the loss in this case is at least fifteen per cent., and often much more. It is a general impression among photographers, that the manufacture of a pure article of the nitrate is attended with much difficulty; this is, however, not the case, unless the result is to be in large crystals, when many precautions are necessary. If, however, the photographer makes the nitrate only for his own consumption, the fine *snow crystals* obtained by rapidly evaporating a solution of the nitrate are even preferable, in many respects, to the well crystalized salt. The chief point to be observed is to obtain a pure and neutral salt, for which the following formula answers admirably. Only the purest metallic silver, as that obtained from the chloride or well refined, should be used for the manufacture of the nitrate. Place the well granulated metal in a large evaporating dish, and upon each ounce pour four ounces of the following mixture:

Pure nitric acid - - - -	4 ounces.
Water - - - - - -	5 "

Place the dish upon a moderately hot stove, and raise the mixture gradually to the boiling point, and let it simmer for about thirty minutes, or until the generation of red vapors and all violent chemical action has ceased; then remove it from the fire, dilute it with an equal amount of water, and filter the liquid through a wad of cotton placed in the neck of a funnel. If it should not be clear at first, it must be filtered again, then poured off in an evaporating dish, placed upon a stove and evaporated to dryness. When the liquid has all evaporated, the remaining mass of white crystals should be exposed to heat until they begin to assume a grayish brown color, caused by the liberation of oxide of silver, the dish must then be quickly removed from the fire, allowed to cool, and the gray mass then dissolved perfectly in distilled water. Finally, filter the murky liquid and dilute it to the proper strength for the sensitizing bath, or evaporate the solution to dryness and preserve the white crystals in a well stoppered glass bottle.

To Make Pure Chloride of Gold.

Place the metallic gold in a small evaporating dish or beaker glass, and upon each pennyweight pour two to three drachms of a mixture of the following acids:

Nitric acid - - - -	3 ounces.
Hydrochloric acid - - -	2 "

Place the mixture upon the stove until complete solution takes place; then dilute with some distilled water and filter the yellow liquid until perfectly clear; then place it in a small evaporating dish and expose it to a gentle heat until it has boiled off to thorough dryness. To obtain the salt of great purity, it may be once more dissolved in a little distilled water and again crystalized; but this second crystalization is hardly necessary. The well dried mass of yellow crystals should be rapidly removed from the dish and placed in perfectly dry and warmed bottles.

To Make Auro-Chloride of Sodium.

Place the gold in the form of powder or otherwise in a small porcelain dish as before, and pour upon it the same amount of acid as in the last.

When the liquid has nearly evaporated to dryness (the second time), rapidly and thoroughly stir it in a quantity of clean common salt to equal the weight of the metallic gold used, and let the mass then crystalize by evaporating the liquid to dryness. Preserve the salt in well stoppered bottles as it rapidly absorbs moisture from the air. Or the solution of gold may be used direct without evaporating; but the solution should have been already once boiled down. Both the nitrate of silver and the chloride of gold prepared by these processes are of great purity. The nitrate of silver is rendered perfectly neutral by exposing the crystalized mass to heat, until a small quantity of the oxide of silver is liberated, and the chloride of gold by double crystalization.

Another Way of Making Pure Chloride of Gold.

The gold should be placed in a flask or glass and covered with an ounce or two of distilled water.

Then chlorine gas is led into the water, which, soon

becoming charged with it, rapidly attacks and dissolves the gold.

The yellow solution thus resulting must be evaporated to perfect dryness in a suitable vessel and treated precisely like the last.

To generate chlorine a small glass flask should be provided with a tight fitting cork, into which is inserted a bent glass tube or India-rubber gas tube. Into this flask place some black oxide of manganese, and pour upon it enough muriatic acid to make it of the consistency of cream. The flask is then placed in boiling water or heated sand, and the chlorine rapidly generates and is absorbed by the water.

To Make Pure Nitrate of Silver from the Oxide.

The simplest process for the manufacture of pure nitrate of silver, and that involving the least apparatus and time, is by the use of the oxide of silver, obtained either by the precipitation of an old solution, or by the action of caustic potash upon the chloride of silver.

Place the well washed and still wet oxide in an evaporating dish of proper size, and add slowly the following mixture of acids:

Chemically pure nitric acid	-	2 ounces.
Distilled water - - -	-	3 "

As soon as violent chemical action takes place, the addition of acid should be stopped, the dish placed upon a sand bath and brought nearly to the boiling point.

If any black powder now remains, more acid should

carefully be added, drop by drop, until the liquid becomes perfectly clear, and on precipitation remains in the solution.

The contents of the dish should then, if necessary, be thrown upon a filter, and the clear filtrate passing through placed in a porcelain dish and rapidly evaporated to dryness. The resulting snowy mass consists of pure nitrate of silver, and should, if an excess of acid has been added, be exposed to a gentle heat until the oxide begins to form, as directed above.

Nitrate from the Carbonate.

When the silver has been precipitated in the form of a carbonate, the well washed, yellowish brown powder should be placed into an evaporating dish, or beaker glass, and treated precisely in the same way as the former.

The results of both of these processes are remarkably pure, provided the directions above given are adhered to.

Testing Nitrate of Silver as to Purity.

The nitrate of silver of commerce, and that sold by most stock dealers throughout the United States, is rarely, if ever, perfectly pure; but, on the contrary, often contains foreign substances, either purposely mixed with it as an adulteration to lesson its cost, or allowed to enter into its composition by carelessness or ignorance.

Some manufacturers, for instance, add nitrate of potash or saltpeter for the purpose, they say, of producing large crystals; and one firm, doing a very heavy business, adds as much as half an ounce to every pound of nitrate. Whether this is done from disinterested motives or not, we leave the photographer to judge.

The operator may readily test the purity of any nitrate of silver as follows:

Dissolve five grains of the nitrate in a drachm of distilled water, and when complete solution has taken place, add chemically pure hydrochloric acid, drop by drop, until it no longer gives a white precipitate. Now place a few drops of the clear liquid above the sediment on a thin strip of platinum, or a watch-glass, and expose it to a gentle heat. If, when the water has boiled off, no residue is left on the surface, the silver is pure; if, on the other hand, much residue is left, it is impure and should be rejected. Of course the acid must be pure and the water distilled, otherwise they both have residues, which would mislead the examiner.

One valuable hint we can give, that is *never buy cheap nitrate of silver*, as it is invariably impure, the profits on a pure article being so small, that only the largest houses can sell it with a profit. No workman can do good work without proper tools, and chemicals may be aptly termed the tools of the photographer. This is the failing of many of our operators. They work with cheap and inferior chemicals, and many do not even know their composition, to say nothing of their reaction mutually upon each other. Without a thorough knowledge of photographic chemistry, the operator cannot possibly arrive at that standard which may be truly termed ART PHOTOGRAPHY.

WEIGHTS AND MEASURES.

The following tables of the corresponding values of French and English weights and measures will be found very useful for reference, as the decimal system of the French, owing to its greater convenience for calculations, is extensively adopted in chemical works.

The *unit* of the French weight is the *gramme*, which is the weight of the hundreth part of a cubic metre of distilled water, at the temperature of melting ice. Below are some comparative tables, which will serve to give every explanation.

		Grammes.		Troy Grains.
Milligramme	=	·001	=	·01543
Centigramme	=	·01	=	·15434
Decigramme	=	·1	=	1·5434

Troy Weight.

Troy weights.			French weights.	
1 grain	= 1-24 dwt.	=	0·06477	gramme.
1 pennyweight	= 1-20th of an ounce	=	1·55456	gramme.
1 ounce	= 1-12th of a lb. Troy	=	31·09130	grammes.
1 pound imperial		=	0·3730956	kilogramme.

Avoirdupois Weight.

English.			French.	
1 drachm	= 1-16th of an ounce	=	1·7712	gramme.
1 ounce	= 1-16th of a pound	=	28·3384	grammes.
1 pound or 1 pound imperial		=	0 4534148	kilogramme.
1 cwt.	= 112 pounds	=	50·7824600	kilogrammes.
1 ton	= 20 cwt.	=	1015·6490000	kilogrammes.

		Grammes.		Troy Grains.
Gramme	=	1	=	15·434
Decigramme	=	10	=	154·34
Hectogramme	=	100	=	1543·4
Kilogramme	=	1000	=	15434
Myriagramme	=	10000	=	154340

Or:

French.	English.
1 gramme	= 15.438 grains Troy = 0·643 dwts. = 0·03216 oz. Troy.
1 kilogramme	= 2·68027 lbs. = 2 lbs. 8 oz. 3 dwts. 6 grs. Troy wt.
1 kilogramme	= 2·20548 lbs. = 2 lbs. 3 oz. 4 4-5 drs. Avoirdupois.
1 myriagramme	= 22·0485 lbs. Avoirdupois.
1 quintal	= 1 cwt. 3 qrs. 25 lbs.

The unit of superficial measure is "the *are*, a surface of ten metres each way, or 100 square metres. The unit of measures of capacity is the *litre*, a vessel containing the cube of a tenth part of the metre, and equivalent to 0·220097 parts of the British imperial gallon. The standard temperature is 32° F. All the divisions and multiples of the units are decimal."

Measures of Length.

Myriametre	=	10,000	metres.
Kilometre	=	1000	"
Hectometre	=	100	"
Metre	=	1	"
Decimetre	=	0·1	"
Centimetre	=	0·01	"
Millimetre	=	0·001	"

Measures of Surface.

Hectare	=	1000	sq. metres.
Are	=	100	"
Centiare	=	1	"

Measures of Capacity.

Kilolitre	=	1000	litres.
Hectolitre	=	100	"
Decalitre	=	10	"
Litre	=	1	litre.
Decilitre	=	0·1	"
Centilitre	=	0·01	"

The unit of the solid measure is the *stere*, or cube of the metre, equal to 35·31658 English cubic feet.

Measures of Length. Long Measure.

English.		French.
1 inch or 1-36th of a yard - -	=	2·539954 centimetres.
1 foot = ⅓ of a yard = 12 inches	=	3·0479449 decimetres.
1 yard = 3 feet - - - -	=	0·91438348 metre.
1 fathom = 2 yards - - -	=	1·82876696 "
1 pole or perch = 5½ yards -	=	5·02911000 metres.
1 furlong = 220 yards - -	=	201·16437000 "
1 mile, or 1760 yards - - -	=	1609·31490000 "

French.		English.		
1 millimetre	=	0·039370		
1 centimetre	=	0·393708		
1 decimetre	=	3·937079		
1 metre	=	39·37079	=	1·093633 yards.
1 decametre	=	393·7079	=	10·936630 "
1 hectometre	=	3937·079	=	109·366300 "
1 kilometre	=	39370·79	=	4 furlongs, 213·633 yards.
1 myriametre	=	393707·9	=	6 miles, 1 furlong, 156·288 yards

French. English.

1 toise = 6·3945 feet = 2·1315 yards = 76·735 inches.

1 aune or ell = 3·893 feet = 46·79 English inches.

Square Measure.

English.		French.
1 sq. yard - - - - - -	=	0·836097 metre carre.
1 rod or pole = 30¼ sq. yards -	=	25·291939 metres carres.
1 rood = 1210 sq. yards - -	=	10·116775 ares.

1 acre = 4840 sq. yards = 40·4671 ares = 0·404671 hectare.

1 metre carre = 1 centiare - - = 1·196033 sq. yard.

1 are = 3·95 English poles - - = 0·98845 rood.

1 hectare = 2 acres, 1 rood, 5 perches = 2·473614 acres.

Measures for Liquids.

English.		French.
1 pint or ⅛th of a gallon	=	0·567932 litre.
1 quart or ¼th of a gallon	=	1·135864 litres.
1 imperial gallon	=	4·5434579 "

Dry Measure.

English.				French.	
1 peck	=	2 gallons	=	9·0869159	litres.
1 bushel	=	8 "	=	36·347664	"
1 sack	=	3 bushels	=	1·09043	hectolitre.
1 quarter	=	8 "	=	2·907813	hectolitres.
1 chaldron	=	12 sacks	=	13·08516	"

French.		English.	
1 litre = 1·760773 pints = ·8803865 quarts	=	·2200966	gallons.
1 decilitre - - - - - - -	=	2·2009667	"
1 hectolitre - - - - - - -	=	22·0096670	"

As a greater convenience for common purposes, the French denominations were, in 1812, arranged as follows:

1 toise or 6 feet = 2 metres	=	6·5618334	English	feet.
1 foot or 12 inches = ⅓ metre	=	1·0936389	"	foot.
1 inch or 12 lines - -	=	1·0936389	"	inch.
1 line - - - - -	=	0·0911365	"	"

1 aune or ell = 1 1-5 metre = 3·937 English feet:—Or,
1 aune - - - - - = 47·244 " inches.

1 bushel = ⅛ hectolitre = 762·85 cubic inches.

1 old Paris foot = 1·066 English foot.
1 old Paris inch = 1·066 " inch.
1 old line = 0·0888 " "

The metre, the square metre, and the cubic metre, are the radical standards of the three measures; for there are only three, as solidity and capacity, differently named and used, are the same in reality. From these radical denominations, namely, the *metre* of the lineal measure, the *are* of the superficial measure, the *litre* of measured capacity, and the *stere* of solid or cubic admeasurement, the larger ones are procured by multiplying by ten, and the lower ones by dividing by the same. Thus,

Deca prefixed,	means	10	times.
Hecto,	"	"	100	"
Kilo,	"	"	1000	"
Myria,	"	"	10,000	"

The number is understood to multiply the surface of the solid and not its side. Thus, one decare is ten ares, not a square of ten times the side of an are, and so of the others.

The denominations below the radical ones are expressed by a sort of Latin prefix. Thus,

Deci is one-tenth.
Centi is one-hundreth.
Milli is one-thousandth.

Morfit's Chemical Manipulations.

PART II.

THE

GOLD WORKER'S GUIDE:

A SHORT AND PRACTICAL TREATISE ON

DISCOVERING, ASSAYING, ESTIMATING

AND

TESTING ORES AND MINERALS

CONTAINING

GOLD OR SILVER.

INTRODUCTION.

Inexperienced persons, and miners not having the advantage of a knowledge of chemistry, have often felt that imperative want of a simple, short and practical treatise on gold and silver ores, the way of discovering, testing and assaying them. There is not, however, we believe, a single popular work which treats the subject in a practical manner; but, on the contrary, the now-existing works are replete with high-sounding technical expressions.

Again, what great dismays and crushing blows have inexperienced persons received by mistaking some shining substance for the precious metal, and only after years of patient, secret keeping, discovering that "all is not gold that glitters."

Comstock, in a "Treatise on the Precious Metals," relates the story of an honest farmer, who through long years preserved the secret of a supposed gold mine of inestimable value, which, when he was about to reward himself for his patience, turned out to be nothing but fool's gold or sulphuret of iron.

Another man, of a somewhat speculative turn of mind, once brought a specimen, as the story runs, of a supposed very rich gold ore to a San Francisco assayer. The man on being informed that it was nothing but iron pyrites, and not worth a cent a ton, exclaimed, in the most extreme dismay: "Great heavens! there is an old woman up our way who owns a hill of it, and I married her!"

The greater part of the chapters have been compiled from that incomparable work "Muspratt's Chemistry," and the editor has tried to make it a thoroughly practical treatise and one which the most unsophisticated can understand.

THE AUTHOR.

BROOKLYN, N. Y., May, 1867.

THE

GOLD WORKER'S GUIDE.

CHAPTER I.

SOURCES OF GOLD, ETC.

More gold is now produced in the vast continent of North America than in any other part of the world. In this respect it has now taken the place which formerly belonged to South America, while the latter has sunk into comparative unproductiveness, less perhaps from the absolute exhaustion of the auriferous soils, than from the want of the capital and enterprise to work them successfully. North America was a gold-producing continent long before the discovery of the Californian treasures; but for many years, the only source of this precious metal in that quarter of the globe was the argentiferous veins of Mexico, from which it was extracted along with the silver, as in the Peru mines of South America. At a later period, however, an extensive gold region was discovered in the United States, extending along the Eastern slope of the Appalachian mountains, from the River Rappahannock in Virginia Southward to the River Coosa, an affluent to the Alabama, which flows into the Gulf of Mexico. The metal is found in less quantity Northward along the same mountainous range to the State of Maine, and even extending into Canada, where a search for profitable workings has lately been prosecuted with some vigor. The existence of spangles and pepitas of gold, in several rivers of the East of Canada, has been fully established; and at the Great Exhibition honorable mention was made of the Chaudiere Mining Company, who exhibited pepitas of native gold, collected in the washings of those streams. But the States of Virginia, North and South Carolina, and Georgia, afford the most productive deposits. In these, as in those of Brazil and Columbia, the auriferous ores are chiefly pyritical; much of the gold is extracted by amalgamation, after stamping under water.

But all preceding gold discoveries in America, or in any other part of the world, were eclipsed by those that have been made within the past few years in California and Australia. These recent discoveries have produced quite a revolution in the annual production of gold, the effects of which, though already powerfully felt in the new impulse given to emigration and commerce, are only beginning to be developed, and must produce the most important results in the future history of the world. This may be inferred from the fact, that the gatherings of the precious metal, reckoning the average produce of all parts of the new and old world for a series of years previous to 1847, did not amount to the annual value of five millions sterling, whereas the amount now exceeds thirty millions per annum.

The first of these recent discoveries was made in 1847, when California, a hitherto-neglected and little-known region, lying at the most remote South-western limit of North America, rose into sudden importance, as the El Dorado of the new world. The gold region, properly so called, occupies the Northern part of California, commencing near the mouth of the Sacramento River, in latitude 39° North, and longitude 122½° West, to the North-east of the bay and town of San Francisco, from which it extends South and North. At this point two rivers unite and discharge themselves into the sea—the Sacramento flowing from the North, along a valley formed by mountain ranges, and the San Joaquim, flowing from the South, along a similar valley, enclosed on one side by the Rocky Mountains, and on the other by the mountainous ridge which protects the Western coast. It was on the property of an intelligent Swiss emigrant, Captain Suter, who had become a wealthy settler on the banks of the Sacramento, that the first traces of gold were discovered in September, 1847. This happened in the course of the erection of saw-mills on the estate, when Mr. Marshall, the contractor for the building of these, observed glitering particles in the sand of the mill-race, which were ascertained to be gold; and on making further researches, it was found that the precious metal was very extensively diffused in the bed of the stream. The discovery soon became known to the work-people, by whom the intelligence was conveyed to San Francisco, and in no long time, the whole population of the little town, and the scattered and scanty settlers in the neighborhood, abandoned their dwellings and occupations to engage in the exciting search. The supply exceeded the most exaggerated accounts that had been given; new and richer localities were discovered; the gold was found in the beds of various streams flowing into the Sacramento; in the mud of the river itself; in the channels of old water-courses, and along the sides of the hills. The intelligence rapidly spread to the neighboring countries—to Mexico, to South America, to the United States, and thence to Europe. People began to flock from all quarters to the once neglected, but now coveted region; it became, in a few short months, the scene of a considerable population, instead of a few scattered tribes of Indians; miners or diggers in parties spread themselves over the face of the country; and San Francisco, from a mere village, grew up into a place of wealth, importance, and stirring activity. It was at first very generally thought that the supply would soon fail, but this anticipation has proved to be unfounded; and while, on the one hand, the amount of the precious metal disseminated in the rocks and soil appears to be almost inexhaustible—on the other hand, the constant increase in the number of miners, combined with the improved apparatus and methods of working, seems to have hitherto resulted in a steadily increasing annual produce, until within the last few years, when it seems to have been nearly stationary.

Professor Blake, who minutely examined the auriferous regions of California, states that, with the exception of the diluvial strata, the whole geological formation of the Sierra range, through which flow the principal rivers, consists of igneous and metamorphic rocks. The former are mostly porphyritic in the lower hills, whilst higher up trachytic rocks are most frequently met with. The metamorphic rocks consist of micaceous schists, slates both talcose and micaceous, metamorphic sandstones and limestones, with occasional beds of conglomerate. In that part of the country which he examined, the extent of the diluvial deposits was commensurate, or nearly so, with that of the gold-bearing region. They are found in a belt of land from thirty to sixty miles broad, and running parallel with the axis of the range. These diluvial deposits are met with toward the lower hills of the Sierra, extending frequently some miles into the plain. The elements of which they are composed differ considerably in various localities, although there are many points of resemblance through the whole series. In the lower valleys and flats, between the ranges of the lower hills, they appear to consist of beds of gravel, containing occasional boulders of quartz and the harder rocks. On the elevated flats, higher up in the mountains, the surface of these deposits is generally covered by a reddish loam, mixed with small gravel; whilst, reposing on the bed rock, and a few inches above it, is found a stratum containing large boulders and gravel, the boulders being principally quartz. At other points, the whole series consists of conglomerates and soft friable sandstone. Where the deposits are found extending over a large surface on the elevated flats, gold is always met with, generally diffused through the gravel immediately above the rock on which they rest, which yields from fifteen to forty cents to the hundred pounds of earth. There are parts where acres of these deposits have been turned up, in which the gravel never contains less than fifteen cents to the hundred pounds, and generally more. In the valleys in the lower hills, and even on the plains to the west of them, where they are extended over vast tracts of country, these deposits are still auriferous, the gold being very generally diffused, and found in greater quantities the deeper they are worked; but sometimes they will not pay for working, owing to the distance from water. In one place where water could be readily obtained, a portion of these deposits, situated to the west of the lower hills, was found to yield from five to thirty cents to one hundred pounds of earth, through an extent of one hundred and fifty acres, the soil being found richer the deeper it was worked.

At a spot which was appropriately named Mount Ophir, the auriferous soil was described as soft clay and slate, saturated with gold in small particles and large lumps. This treasure was found from ten to thirty feet below the surface, and seven Mexicans who made the discovery, and kept their secret eight days, made in that short time two hundred and seventeen thousand dollars.

Other searchers, from a shaft twenty feet deep, obtained the soft clayey slate in buckets, and found from eight to twelve dollars' worth in each bucket. In many cases considerable nuggets are met with, but no accounts speak of very large masses of gold having been found in California. The total produce from this region, down to the end of 1855, was estimated at sixty-four million pounds sterling; and, latterly, it has averaged about fourteen million pounds per annum.

METHOD OF EXAMINING AURIFEROUS DEPOSITS.

As a general rule, the rocks in the district to be examined for gold, should be either granitic, porphyritic, or quartzose, although it is found in other formations, and particularly in clay-slate. The auriferous quartz is often stained of a rusty brown color, from the presence of peroxide of iron, and in many instances presents a cellular or honeycombed appearance. The points to be most carefully examined are the sands of the rivers and streams, or old water-courses, as well as the particles of disintegrated rock which often accumulate in the eddies of ravines formed on the sides of hills by the action of water during great floods. The sections of rock exposed by this action must also be examined with a view to the discovery of veins of auriferous quartz, from which fragments are broken off and afterwards carefully assayed.

The method of conducting a systematic assay of gold ores and alloys will be fully explained afterwards. It is a somewhat tedious and difficult process, requiring considerable experience and a regular assortment of apparatus. But even in the absence of these, a tolerably correct estimate of the amount of gold present, may be readily arrived at by the following simple method.

The fragment of rock, supposing the ore to be quartz, is first pounded very fine and sifted, a portion of the sand or powder thus obtained is washed in a shallow pan, and, as the gold sinks, the lighter portions of the substance are allowed to float off. The greater part of the gold is thus left in the angles of the pan; and by adding more of the powder, and repeating the same process, a further portion is obtained. When the bulk of powder, with which the gold is mixed, is thus reduced to a manageable quantity, mercury is added to the mass, and forms with the gold an amalgam, which is afterwards heated in an iron retort, to expel the mercury. In this way the proportion of gold contained in a specimen of rock may be ascertained with considerable exactness. The sands brought down by rivers are examined in much the same manner, but do not require the previous pounding; and a fair estimate of their value may be generally formed without the amalgamating process. Sometimes, indeed, the gold may be present in considerable quantity, although in a state of division so minute as not to be readily perceived by the unassisted eye, and, therefore, in examining the earthy residuum, a small magnifying lens will be found of great use.

It is generally conceded that the sand of any river is worth working for the gold it contains, provided it will yield twenty-four grains to the hundred weight; but the sands of the African rivers often yield sixty grains in not more than five pounds weight, which is in the proportion of more than fifty times as much, while the Australian rivers have been known to yield considerably more.

SUBSTANCES OFTEN MISTAKEN FOR GOLD.

Though gold, in its separate and pure state, is readily distinguishable from other metals and minerals, by its color, softness, and high specific gravity, and insolubility in the simple acids, yet in the mixed state in which it is generally found, it requires an amount of chemical skill to apply these tests, and hence, where glittering particles of other minerals appear, they are often mistaken for gold, even by persons who are not entirely ignorant of its characteristics. The substances which most generally lead to this mistake are iron pyrites, copper pyrites and yellow mica.

Common iron pyrites, or *bisulphide of iron*, which is more frequently mistaken for gold than any other substance, occurs in small cubical crystals, in veins dissemminated in the various slate rocks, and in the coal measures. It is of different shades of brass-yellow, and often in fact contains minute traces of gold, though seldom a sufficient amount of that metal to render its extraction profitable. It may be readily distinguished from gold by the application of the magnet, as well as by the following characters : first, instead of flattening like gold under the hammer, it is extremely brittle, and, therefore, readily broken ; second, its weight or specific gravity is only about one-fourth that of gold ; and, lastly, when heated with nitric acid it is dissolved with evolution of copious red fumes, whilst gold, when so treated, remains unaffected. It was only recently the Editor was consulted by a gentleman who imagined he had discovered a gold mine in Ireland. The sample brought turned out to be a fine specimen of sulphide of iron embedded in quartz.

Copper pyrites, or *yellow copper ore*, the second mineral which is frequently mistaken for gold, is a ferrosulphide of copper, and may be considered as a compound of two equivalents of sulphide of iron, and one equivalent of sulphide of copper. This is the ore from which the largest proportion of the copper of commerce is derived. It occurs in a variety of forms, its primitive crystal being the regular tetrahedron. It is formed in lodes or veins, which usually occur either in granite, grauwacke, or clay-slate, and has a strong metallic lustre, and deep brass-yellow color. It may readily be distinguished from gold by the circumstance, that when heated on a piece of charcoal before the blow-pipe it loses this yellow color and fuses into a dull black globule which, from the presence of the iron, is magnetic. If mixed with corbonate of soda and a little borax, it yields, when similarly treated, in skillful hands, a button of metallic copper. But an easier method to determine the presence of copper is to pulverize the ore in an iron mortar, or with a heavy hammer, dissolving the powder thus obtained into nitric acid, and evaporating the solution nearly to dryness; water is then added, and afterward ammonia in excess, when, if copper be present, the liquor assumes a rich purplish color, an unmistakeable sign of the presence of copper.

Another method for determining the presence both of iron and copper, is to take some of the scoria left after putting this in a blowpipe, and, putting this in a test-glass, to pour over it a few drops of hydrochloric acid, when an effervescing solution will be obtained. A little of this liquid is then transferred to another glass—the one to be treated for iron, the other for copper. Into one of the glasses introduce a few drops of ferro-cyanide of potassium, and liquid ammonia into the other. If iron is present, the liquid in the glass to which the ferro-cyanide has been added will become blue, the iron combining with the ferro-cyanogen, and producing Prussian blue. If copper is present in the same glass, but no iron, the liquid will become of a reddish-brown tint, ferro-cyanide of copper being formed. If iron and copper are present together, the two metallic precipitates will be the result, and a purple tint arises from the mixing of the red and blue colors. The liquid in the glass to which ammonia was added will be changed to a brownish tint if iron is present, and to a fine blue if copper is there.

Mica, the third substance often mistaken for gold, is one of the constituents of gneiss, granite, and mica slate, and gives to the former its lamellar structure.

The specific gravity of mica never exceeds 3·00; and this circumstance, together with its foliated structure, is quite sufficient to distinguish it from gold, which it somewhat resembles in color; but even in this latter particular the microscope will dispel the illusion.

CHARACTERS OF NATIVE GOLD.

Gold, as already stated, almost always occurs in the metallic state, generally in small grains or scales known as *gold dust,* sometimes in particles so minute as to be invisible, but occasionally in pieces of considerable weight, termed *nuggets.* It appears doubtful, indeed, whether, when this metal occurs in pyrites, it exists in every instance in metallic particles, or whether, in some cases at least, it may not be present in combination with sulphur. The auriferous pyrites, as Dumas remarks, contain the gold disseminated through their mass in such small quantities, that it is almost always impossible to ascertain, even with the aid of the microscope, in what state the precious metal exists. As a preliminary roasting of this auriferous ore is generally useful, with a view to the subsequent amalgamation, a doubt may be entertained whether it is really present in the metallic state, though this has been generally assumed hitherto. But Dumas thinks that the powerful electro-negative tendency of sulphide of gold affords a strong presumption in favor of the hypothesis, that this metal may exist partly, or even entirely, under the form of a double sulphide, in iron and copper pyrites, et cetera. Brogniart observes, that it is chiefly in its association with those sulphides, as also with galena or sulphide of lead, blende, or sulphide of zinc, and mispicked or arsenical pyrites—a sulphide of arsenic and iron—that the gold becomes invisible to the eye, a circumstance which is justly regarded as strongly confirming the opinion expressed by Dumas. The other minerals with which it is fouud associated are grey cobalt, lithoidal manganese, native tellurium, malachite, sulphide of silver, red silver and sulphide of antimony.

COMPOSITION OF NATIVE GOLD.

Native gold is never quite pure, being almost invariably alloyed with silver, and containing frequently small portions of copper and iron. In Siberia it is often associated with platinum, and in the Gongo Soco mines in Brazil, an alloy of gold and palladium of a pale yellow color is sometimes found. In Columbia a somewhat similar mixture is procured, in which the palladium is replaced by another rare metal called rhodium. In Hungary it is met with in combination with tellurium and other elements. The specific gravity of native gold varies from 13·3 to 18·5.

The proportion of silver, the principal ingredient which is found in combination with gold, varies from one to fifty per cent., and not only differs greatly in specimens of native gold obtained from diverse regions, but even to a certain extent in specimens from the same auriferous district. In general, however, the composition of the gold of the same district is remarkably constant; so much so that the knowledge of the locality whence it is derived, is often sufficient to enable the experienced assayist to guess pretty nearly the quantity of pure gold which the compound contains. The subjoined table presents the composition of native gold from various parts of the world :

Locality.	Authority.	Gold.	Silver.	Cop'r.	Iron.
SIBERIA AND URAL.					
Schabrowski,	Rose,	98·96	0·16	0·35	—
Boruschka,	Rose,	94·41	5·23	0·36	
Katherinenburg,	Rose,	92·80	7·02	0·06	0·08
Gozuschka,	Rose,	83·85	16·15	—	—
Ural,	Awdejew,	70·86	28·30	0·84	
AFRICA.					
Anamaboc,	T. H. Henry,	98·06	1·39	0·15	
Anamaboc,	T. H. Henry,	88·25	11·17	0·10	0·36
AMERICA.					Palla-dium.
North of Brazil,	Rivot,	91·0	8·7	0·3	
Gongo Soco,	T. H. Henry,	83·36	6·44	0·50	3·58
Ojas Anchas,	Boussingault,	84·5	15·5	—	—
Santa Rosa de Osos,	Boussingault,	82·4	15·5	—	—
Marmato,	Boussingault,	73·45	26·48	—	—
Titiribi, Columbia,	Rose,	76·41	23·12	0·03	—
California,	T. H. Henry,	90·12	9·01	0·87	Iron.
California,	T. H. Henry,	86·57	12·33	0·29	0·24
Canada,	T. H. Henry,	90·38	9·53	—	—
AUSTRALIA.					
Bathurst,	T. H. Henry,	95·68	3·92	—	0·16
EUROPE.					
Transylvania,	Rose,	60·49	38·74	0·77	
Wicklow,	Mallett,	92.32	6·17	—	0·78

From this table it will be seen that the gold both of California and Australia contains silver, but that the specimen from the latter was remarkably pure; and such, indeed, is the general character of Australian gold. Neither platinum nor palladium, nor any trace of the metals of that class is found in the newly discovered gold regions. There is, however, a trace of iron in the Australian specimen and a small proportion of both copper and iron in those from California.

PROPERTIES OF PURE GOLD.

Pure gold is of a rich reddish-yellow color and high metallic luster; in the pulverulent state it is brown and dull, but acquires, as has been stated, the metallic luster by pressure. The specific gravity of fused gold is 19·2; of hammered gold from 19·3 to 19·4. In this respect, therefore, it stands second only to platinum, of which the specific gravity is about 21·5. Finely divided gold, precipitated by sulphate of iron, was found to vary in density from 19·55 to 20·72; and when precipitated by oxalic acid its density was 19·49. Its equivalent is 197. In a pure state it is softer than silver and nearly as soft as lead; but its tenacity is so great that it may be drawn out into very fine wire; and such is its malleability that it may be hammered out into leaves only one three hundred and seventy thousandth of an inch in thickness. A single grain may be extended over 56·75 square inches of surface, or drawn out into a wire five hundred feet long. Reamur, by rolling out a fine gilt silver wire, reduced the coating of gold to the twelve millionth of an inch in thickness, and the surface appeared to be perfect when viewed under the microscope.

Gold does not combine directly with oxygen, and, therefore, suffers no change by exposure to air and moisture at any temperature—not even by being kept in a state of fusion in open vessels. It is not attacked by the mineral, or any of the simple acids, except selenic, and this by the aid of heat. The alkalies do not affect it, and hence a crucible of gold is a valuable instrument in the analysis of minerals which require fusion with the caustic alkalies. It is not acted on by sulphur and, consequently, sulphide of hydrogen is not decomposed by it, as in the case of silver. Iodine has only a weak action upon it, but bromine and chlorine attack it easily at ordinary temperatures; and it is dissolved by any substance which liberates chlorine. It is therefore dissolved by hydrochloric acid, if binoxide of maganese, chromic acid, etc., be added thereto. Its usual solvent is that already stated—a mixture of one part of nitric acid and four parts of hydrochloric acid. *The proper solvent of gold is nascent chlorine*, which is eliminated by the mutual action of the mixed acids.

Gold is one of the most perfect conductors, both of heat and electricity. It fuses at a bright red or a white heat, the temperature of which has been estimated at 2016° Fahr. It is therefore less fusible than silver or copper, the former fusing at 1873°, and the latter at 1996°. In fusion it exhibits a bluish-green color. It is not sensibly volatile in the strongest heat of a blast furnace; but in the focus of a large convex lens, in the intense heat of the oxyhydrogen jet, or under the influence of a powerful electric discharge, a gold wire is dispersed in vapor; and if, in the latter case, the wire be placed just above the surface of a sheet of paper, the course of the discharge is marked by a broad, dark, purple stain, produced by the finely-divided gold. If, instead of the sheet of white paper, a plate of polished silver be employed, it is traversed by a brightly-gilded line, which is firmly attached to its surface. When a globule of gold is placed between the terminal charcoal points of a powerful voltaic battery, it enters almost into fusion, and gives off abundant metallic fumes.

Gold contracts on cooling, and cannot be advantageously employed for castings, as it shrinks greatly at the moment of solidifying. Graham states that it cannot be obtained in crystals by cooling; but according to other authorities, when large quantities of gold have been fused, and are then allowed to cool slowly, cubes more or less modified on their edges and angles are frequently the result. It has been shown that native gold affords numerous well defined crystals belonging to the cubic system, and that of these the greaier uumber, if not all, is affected by the faces of the regular octahedron.

CHAPTER II.

CHEMICAL EXAMINATION OF GOLD ORES.

A RUDE method of ascertaining the presence of gold in crushed quartz or earthy ore, by washing with the hand-basin, has been described; but gold is often present in the matrix in grains or particles so minute that it cannot be detected by the eye, and sometimes it is quite disguised by admixture with other metals or minerals. Many cases occur, however, in which it is of great importance to ascertain the presence of gold, not only as a first step in the examination of suspected auriferous ores, but likewise in the assaying of alloys of gold with different metals. In commencing mining operations, the fact of the *presence of gotd* in the ore in any appreciable quantity, is the first point to be determined; and if this be decided in the affirmative, the next point is to determine the *proportion or quantity of gold* contained. The first process is termed *testing for gold*, or the *qualttative examination* of a suspected mineral; the second is the *quantitative examination*, or *assaying* process.

Sulphate of iron, protochloride of tin, and oxalic acid are the tests or reagents pre-eminently employed in seeking to determine the presence of gold.

Before any of these tests can be applied, the substance supposed to contain the gold must be brought into a state of solution; and this can only be done by means of the mixture of nitric and hydrochloric acids already mentioned. If the substance consist of earthy or quartzose matter, this must be reduced to powder by trituration in a mortar before it is subjected to the action of the solvent; but if the matter under examination be simply a metallic alloy, it can be dissolved without any previous preparation. An excess of acid should be avoided, and for this purpose some carbonate of soda should be added. When the solution is effected, the liquid should be evaporated to about one-eighth of its original bulk, and then diluted with three or four ounces of water. The action of the reagents is as follows:

1. *Sulphate of Iron or Green Vitriol.*

If a few crystals of this salt be dissolved in distilled water, and dropped into the suspected solution, the result is the precipitation of the gold if any be present, in the form of a dark-brown powder, which is metallic gold in a very fine state of division, as already described in connection with the laboratory process for preparing pure gold. If the solution has been mixed with a considerable quantity of water, the liquid, on the addition of the green vitriol, is colored brown by reflected, and blue or a dingy green, by transmitted light; and this is obvious even when forty thousand parts of the menstruum are present to one part of gold. If the liquid amount to eighty thousand parts it is colored sky-blue; with one hundred and sixty thousand parts it becomes violet; with three hundred and twenty thousand parts of liquid, the violet tint is still very obvious; but with six hundred and forty thousand parts, it is with difficulty perceived.

The deposit formed when sulphate of iron is added, may be corroboratively proved to be gold, by its being insoluble in nitric acid, but readily soluble in aqua regia.

2. *Protochloride of Tin.*

If to another portion of the nitro-hydrochloric solution be added a small quantity of a solution of protochloride of tin—commonly known as *salts of tin*—there will be immediately produced, if any gold is present, a dark brownish-purple precipitate known as *purple of Cassius.* This substance is used in enamel and porcelain painting, and also for tinging glass of a fine red color, in connection with which applications it will be noticed afterwards. Its color, though not a brilliant purple, but rather a reddish-brown, is characteristic, and after being once seen is not likely to be mistaken. Its appearance, when the chloride of tin is added to the liquid, affords an infallible proof of the presence of gold, for a very minute portion of that metal gives a manifest reaction when this test is employed.

When the first test—sulphate of iron—has been applied, and its evidence corroborated by the solution of the precipitated gold powder in aqua regia, the protochloride of tin may be employed to produce the purple of Cassius in this solution also.

3. *Oxalic Acid.*

This substance, either in crystals, or dissolved in water, causes, when added to the nitro-hydrochloric solution, the precipitation of any gold that may be present, in the form of a brown or greenish-black powder, in the same manner as the sulphate of iron; but the precipitation does not occur so rapidly. Indeed, it requires not less than forty-eight hours for the whole of the gold to be thrown down by the oxalic acid, unless heat is applied, by which the process is accelerated. The precipitate is pulverulent gold, and may be tested, as in the first case, by its insolubility in nitric acid, while readily dissolving in aqua regia. A crystal of oxalic acid, wetted with a solution of gold, becomes soon covered with a thin film of the metal, having its distinctive color and luster.

These are the tests most easily employed by persons unaccustomed to chemical manipulation. At the same time, it may be useful to know the reactions of gold with various other substances, the most important of which, including the three already mentioned, are summarily tabulated as follows by Dr. Lyon Playfair, in a lecture delivered by that chemist at the Museum of Practical Geology:

TESTS FOR GOLD.

Tests or reagents.	Results.
Sulphate of iron, . . .	In acid solution, brown precipitate; if very dilute solution, only a blue coloring.
Protochloride of Tin, . .	In dilute solution, a purple-red coloring; when strong, almost brown precipitate.

Tests or Reagents.	Results.
Metallic zinc, ...	Precipitates metallic gold as a voluminous brown precipitate.
Potassa in excess, . . .	No precipitate; after some time a green coloring and slight precipitate.
Ammonia, ...	Yellow precipitate—fulminating gold.
Carbonate of soda, or carbonate of potassa, . . .	No precipitate in cold solutions, but when heated, voluminous precipitate like oxide of iron.
Bicarbonates of soda or potassa	No precipitate.
Carbonate of ammonia, . .	Behaves like ammonia, carbonic acid being evolved.
Oxalic acid, ...	Dark, greenish-black precipitate, more quickly produced by heat.
Tartaric acid and soda, . .	Dark precipitate when boiled.
Sulphide of ammonium and Sulphide of hydrogen, . .	Dark brown or black precipitate.

These reactions are so characteristic that it is impossible to mistake gold for any other metal. At the same time it may be stated that the protochloride of tin is the most infallible test, and is indeed quite conclusive. It has the advantage of being more delicate than the others—that is to say, it will indicate the presence of a smaller quantity of gold than any other reagent, not excepting the protosulphate of iron or the oxalic acid.

TEST FOR METALS USUALLY ASSOCIATED WITH GOLD.

In examining an ore or alloy supposed to contain gold, it is often an object of great importance to determine the nature of the metals with which the gold is associated. These may be of high value on their own account, and, according to the quantities in which they are present, may greatly affect the value of the ore or mineral in question. To enter fully into this subject in connection with the rarer metals that are found associated with gold, would be of little practical use. The Editor will therefore confine his attention to the means of detecting and distinguishing those of most common occurrence—copper, silver, and platinum.

1.—*Copper.*

It has been shown that copper is almost always associated with gold, even in quartz, and that copper pyrites is one of those substances frequently mistaken for gold. When dissolved in acids, however, it gives characteristic actions, which render its presence easily distinguishable. One of the readiest tests is to introduce into the solution a piece of clean iron, when, if copper be present, it will be deposited on the iron in the metallic state. This experiment, remarks Dr. Lyon Playfair, apparently showing the conversion of iron into copper, deceived the alchemists in their researches, and gave much support to the idea that one metal may be transmuted into another. The action, depends, however, upon a simple exchange, the iron going into the solution in proportion as the copper goes out. Again, when ammonia is added in excess to a solution in which copper exists, it communicates to the liquid a rich deep blue color. Ferrocyanide of potassium produces, with copper, a brownish-red precipitate, even when the metal is present in very small quantity. Carbonate of soda precipitates copper from its hot solutions in the form of an apple-green compound, which is a carbonate of copper known, when artificially formed, as *verditer*, and when it occurs native, as *malachite*. Copper ore in the latter form exists abundantly in Australia—not, indeed, mixed with the gold, but constituting valuable mines, from which the ore is sent over to England, with great profit, to be smelted in South Wales.

2. *Silver.*

It has been shown that gold appears to be invariably alloyed with this metal, sometimes to a very large amount. In its separate state it is readily distinguished, not only by its white color, but also by its specific gravity, which is only 10·4, or about one-half that of gold. It may be useful to state, that when in a very fine state of division it is of a dark grey color. It may be easily recognized by its chemical behavior to reagents, in which respect it differs from gold by solubility in nitric acid at all temperatures, and in boiling or heated sulphuric acid. On the contrary, with hydrochloric acid, it forms a white curdy precipitate, which is the chloride of silver. If the nitric acid employed to dissolve it contain the least hydrochloric acid, the solution will become turbid by the formation of the chloride. Hence, when a mineral containing gold and silver is submitted to the action of aqua regia —nitro-hydrochloric acid—the appearance of this white precipitate will immediately indicate the presence of the latter metal. The chloride of silver is soluble in ammonia, and may thus be distinguished from many other white precipitates; or it may be further tested by putting the precipitate into a crucible with carbonate of soda, and exposing the mixture to a strong red heat, when a button of pure silver will be obtained. By careful manipulation the amount of silver present may be accurately determined in this manner. If the mineral containing the silver be dissolved in oil of vitriol, it is readily detected by inserting a few fragments of copper, which causes the precipitation of the silver in a pulverulent state.

From the fact that silver, in greater or less proportion, is uniformly associated with gold in the native state, it may be useful to compare the following reactions with those given in the preceding page, for the metal of higher value:

TESTS FOR SILVER.

Tests or Reagents.	Results.
Potassa,	Brown precipitate, becomes black on boiling.
Ammonia,	Brown precipitate, soluble in excess of ammonia.
Carbonate of soda or potassa,	White precipitate, soluble in ammonia.
Carbonate of ammonia,	White precipitate, soluble in excess of reagent.
Phosphate of soda,	Yellow precipitate, soluble in ammonia.
Oxalic acid,	In neutral solutions, white precipitate.
Sulphide of hydrogen and sulphide of ammonium,	Black precipitate.
Hydrochloric acid,	White curdy precipitate, soluble in ammonia.
Zinc or copper,	Precipitates white metallic silver.
Sulphate of iron,	In neutral solutions, white metallic silver.

3. *Platinum.*

This is another metal frequently associated with gold; and as it is one of the noble metals, and ranks in price between silver and gold, an ore which contains it in any considerable quantity is of great value. The specific gravity of platinum is about 21·5, and is, therefore, greater than that of gold. In short, platinum is the heaviest of all known substances. It is of a light steel-grey color, less ductile than gold or silver, but more tenacious, and will support greater weights on equal thicknesses of wire than any metal except iron or copper. It is distinguished from gold not only by its color but also by its extreme difficulty of fusion; it does not melt by itself in the highest heat of a furnace, but softens sufficiently to admit of forging and welding, and in the arc of flame of the voltaic current, or before the oxyhydrogen blowpipe, it admits of being fused in considerable masses. On the other hand, it resembles gold, not only in its high specific gravity, as already stated, but also in the fact that it resists the action of all the simple acids, and is soluble only in aqua regia. This circumstance, together with its great infusibility, renders it importantly useful in many of the arts and indispensable for various purposes in the chemical laboratory. It is the metal universally employed for apparatus which requires to be exposed to high temperature and powerful chemical agencies, without undergoing any change. It is, therefore, very desirable, as Dr. Lyon Playfair has remarked, that those who go to the gold regions should look well for this precious metal, as it is likely to escape the notice of the common observer, from its less glittering appearance.

There are, however, certain chemical reactions by which platinum may be readily distinguished and separated from the gold in solution. Sulphate of iron and oxalic acid, which precipitate gold, do not precipitate platinum. When the latter is dissolved in aqua regia, and the acid neutralized by carbonate of soda, it is deposited as a black powder, if the mixture be boiled with tartaric acid and soda—the ingredients of a Seidlitz powder. Further, the addition of chloride of ammonium and alcohol to a strong solution of platinum, causes the deposition of a yellow crystalline precipitate, which is characteristic of this metal. These and other reactions may be summarily stated as follows:

TESTS FOR PLATINUM.

Tests or Reagents.	Results.
Chloride of potassium or chloride of ammonium, . .	Yellow crystalline precipitate produced readily by the addition of alcohol.
Potassa or ammonia, . .	With chloride of platinum acts like the chlorides of potassium and ammonia.
Carbonates of potassa and ammonia, ...	In chloride solution yellow precipitate.
Carbonate of soda, . . .	No precipitate.
Sulphate of iron, . . .	No precipitate.
Oxalic acid, ...	No precipitate.
Protochloride of tin, . .	Dark-brown red coloring, no precipitate.
Sulphide of hydrogen and other sulphides, . . .	Dark-brown, nearly black precipitate.
Tartaric acid and soda, . .	On boiling, metallic platinum falls as a black powder.
Zinc, ...	Slowly precipitates metallic platinum as black powder.

Application of the Reagents.

Having thus described the properties and characteristic reactions of the different metals which it is desirable to look for as being frequently associated with gold, let it now be assumed that the substance to be examined is a piece of auriferous quartz. This must be first reduced to powder, and then boiled for some time in an earthenware or glass dish with aqua regia. The solution is then diluted with water, passed through a filter and allowed to cool. If any silver be present it will remain in the filter as a white precipitate mixed with the quartz.

To the liquid which has passed through the filter a solution of carbonate of soda is now added, until no more effervescence takes place. This will precipitate all the other metals which may be present, except gold and platinum. These will remain in solution.

The liquid is again filtered, and a solution of oxalic acid is added until it ceases to produce effervescence, and has a sour taste; then boil; if there be any gold present it will be precipitated as a black powder. The platinum, if any be present, will still remain in solution.

Decant or filter the liquid from the gold precipitate, and add to the former protochloride of tin, when a reddish-brown coloring will appear if platinum be present. Or, by boiling with tartaric acid and soda, the platinum will be thrown down as a black precipitate.

It has been stated that if any silver be present it will be found on the first filter, mixed with the quartz. Wash this with ammonia, which, if copper be present, will produce a deep blue tinge. To the solution which comes through, add hydrochloric acid until the smell disappears, and the silver will be thrown down as a white curdy precipitate.

It is evident that other methods and reagents may be adopted. For example, the original solution in aqua regia may be concentrated by evaporation, until it is very much reduced in quantity; then add about three-fourths of its bulk of spirit of wine, and lastly, a saturated solution of chloride of ammonium. By these reagents the platinum will be thrown down as a yellow crystalline precipitate, while the solution filtered from this, and treated with sulphide of iron or boiled with oxalic acid, deposits gold.

By carefully weighing the gold obtained, the amount present in a given quantity of the ore or alloy may be exactly determined; but full details of the different methods of conducting the quantitative examination, including the assaying of gold ores and alloys by the dry process, will, to avoid repetition, be postponed to the close of the article.

CHAPTER III.

QUANTITATIVE ESTIMATION OF GOLD ORES AND ALLOYS.

In proceeding to determine the exact amount of gold present in an ore or alloy, it is obviously necessary to exercise the greatest caution in the sampling. Excellent advice on this point is given by Dr. Percy in his admirable lecture on the metallurgical treatment and assaying of gold ores, delivered at the Museum of Practical Geology. Careless sampling, he remarks, can only mislead; assays of individual specimens may be accurate, but they are worse than useless if the assayer has not operated upon an *average* sample of the ore. He therefore advises the capitalist, to whom prospectuses of gold-mining schemes may be submitted, not to be allured with glittering specimens of gold ore, with assays yielding a high produce, and with the glowing statements of sanguine promoters or enthusiastic adventurers, without having ascertained on good evidence that the samples which are presented to his notice are really *average samples* of the ore, and that something like a continuous supply may reasonably be expected. If such specimens, he adds, do not represent an average, they become what the Cornish miner calls *slocking-stones,* which are at all times enticing and dangerous to the inexperienced and unwary, and never more so than in the case of auriferous ores. The sampling generally devolves upon the miner, but the assayer and metallurgist should likewise understand the business. Assayers of great experience and high integrity, may occasionally, he adds, commit unintentional mistakes. Thus, a few years ago, two small pigs of lead from South America, very rich in silver, were offered for sale. They were assayed by men of very high standing. Portions had been taken from the top and bottom of each pig with a view to obtain a fair average. Dr. Percy had occasion to attempt to verify the report of the assayers. Portions were taken from the same parts of each pig as in the first instance; but the results did not agree with the report, nor did Dr. Percy's assays agree with each other on taking fresh portions. It was therefore certain that the composition was not uniform and that the portions taken for the purpose of assaying in neither cases represented an average. The pigs were accordingly sent again to the same assayers. Each pig was melted separately, and while melted a sample was taken. A second report was given, which differed from the former to the extent of one thousand ounces and upwards to the ton! In the sampling of gold ores most especial care should be taken, as the precious metal exists irregularly diffused through the mass, in particles of very different size, and as minute errors in sampling will ncessarily be greatly multiplied when the quantity of gold per ton is calculated from the assaying of, it may be, five hundred or a thousand *grains* of ore.

1. *Quantitative Estimation by the Touchstone.*

A method of estimating approximately the amount

of gold, not in auriferous ores, but in native gold or artificial alloys, is by the use of the touchstone—a method which may be practiced with considerable advantage when the apparatus and reagents necessary for the carrying out of a complete assay cannot be conveniently procured. It is more especially applicable to the estimation of trinkets or other finished articles, which could not be submitted to the assaying processes, either by the wet or dry method, without destroying them.

The touchstone test essentially consists in rubbing some convenient part of the object to be examined on a smooth piece of black basalt or pottery, which for this reason is termed the *touchstone*, and comparing the marks so formed with those produced by one or other of a series of small bars or needles, consisting of alloys of gold with silver or copper in known proportions. The material commonly employed as the touchstone, and generally known by that name, is a species of quartz, colored dark by bituminous matter, and of which large quantities are found in Saxony, Bohemia and various other localities. Black flint slate will serve the same purpose. The sets of needles or bars may vary from pure gold, through a well-graduated proportion, to equal parts of gold and silver, equal parts of gold and copper, or various mixtures of all three metals in determinate quantities. The fineness of each bar is marked in carats—a mode of valuation which has been already explained.

In proceeding to make an assay by this method, the first streak obtained by rubbing the object to be examined on the touchstone cannot be safely employed to ascertain its composition, if it be a manufactured article, because, by a process previously described, the surface of jewelry is invariably rendered of a higher standard than that of the mass. The object must therefore be passed once or twice over the back or edge of the stone, in order to remove the superficial film of richer metal, before making the streak from which its quality is to be judged. Other streaks are then made successively with two or three of the needles which the assayer, guided by experience, considers to approach nearest in composition to the article under examination. In doing so, he compares not only the color of the streaks made upon the touchstone, but likewise the sensation of roughness, dryness, smoothness, or greasiness, which the texture of the rubbed metals excites when abraided by the stone. When he succeeds in obtaining with one of his needles a streak which is perfectly similar in appearance to that produced by the article which forms the subject of experiment, he then moistens both streaks with nitric acid, which will affect them differently if they be not similar compositions. That which has the least gold will be most affected; on the contrary, if the gold be pure, the streak will remain unaltered. If the actions do not correspond, his experience will enable him to judge in what they differ, and will direct him in selecting another needle to submit to the same comparative test. When one has been found which agrees satisfactorily in all particulars with the metal submitted to examination, the latter is assumed to possess a similar standard of fineness to that which is indicated by the mark on this particular needle.

Nitric acid of specific gravity 1·20 is commonly employed in this operation, with sometimes an addition of about two per cent. of hydrochloric acid. Although the results obtained cannot be relied on where absolute accuracy is required, yet they afford a useful approximation, not only in estimating the value of manufactured articles, which cannot be submitted to a regular assaying process, but also in obtaining that preliminary knowledge of the general composition of an alloy, which is so important to the assayer in proceeding to a detailed analysis. The touchstone is therefore of great use in practiced hands, but it is of little avail in the hands of an inexperienced operator.

2. *Quantitative Estimation by the Wet Method.— Analysis:*

However useful the approximations obtained from the specific gravity of an auriferous ore, or the application of the touchstone to an alloy of gold, it is evident that more exact methods are required to determine with absolute precision the value of an alloy or or ore of this precious metal. Accordingly, there are two methods by which the assaying of gold ores or alloys may be conducted with perfect accuracy, and these are distinguished as the *wet* or *dry* method, according as the agency of liquid solvents, or that of fluxes and fire is employed. For practical purposes, in the determination of gold, the latter process is always adopted, although in the final separation of the silver even the dry method, as now practised, involves the application of a liquid solvent.

The principles of the quantitative estimation of gold

by the humid method have been already explained in detailing the laboratory process for the preparation of pure gold, and in describing the application of the tests or reagents by which its presence is detected. It is evident that any of the qualitative tests which result in precipitating the gold from its solution in aqua regia, such as protosulphide of iron, may be converted into the means of obtaining a quantitative estimation, by simply weighing the amount of ore or alloy on which the experiment is performed, and then weighing the amount of pure gold, carefully washed and dried, which is obtained in the form of a precipitate. It will therefore be unnecessary to enter into minute details on the chemical principles involved.

It may be stated, however, that if the substance to be examined by a natural or artificial alloy, composed chiefly of the pure metal, such as the gold dust of California or Australia, the quantity taken for the analysis should not exceed fifteen grains. A convenient quantity is twelve and a half grains, as it is then only necessary to multiply the result by eight to to obtain the percentage composition. In performing analyses very small quantities are taken, because it is easier to operate upon a few grains than upon a large mass; the effect of the reactions is more rapid; there is less waste of materials; and as accurate results can be obtained with a small as with a large quantity, provided sufficient precaution be taken to operate upon an average sample. If the gold be mixed with earthy or quartzose matter, so much of this should be taken as may be judged, from preliminary experiments or other sources of information, to contain the amount of native gold above mentioned; and this must be triturated in a mortar with great care before subjecting it to the action of the aqua regia—a precaution of less importance in merely testing for gold, but absolutely necessary and essential when it is required to ascertain the exact amount of the precious metal present; for, unless the quartz be reduced to a state of the most minute division, it is evident that much of the gold remaining enveloped in the quartz will escape the action of the acid.

Whatever the precise amount taken, the ore or alloy to be examined must be weighed with extreme accuracy, and then introduced into a Florence flask or any other convenient glass vessel for boiling liquids. Supposing the compound to contain from twelve to fifteen grains of native gold, about an ounce of aqua regia is now poured upon it; the flask is then placed on a retort stand or sand-bath, and the mixture is allowed to digest at a moderate heat for about half an hour. At the end of that time the gold will be completely dissolved, and the silver, if any be present, will be found in the form of an insoluble chloride, mixed with the silica at the bottom. The heat may then be increased under the flask, and the solution boiled off until it is diminished to about an eighth or tenth of its original quantity. At the same time a little hydrochloric acid should be added occasionally for the purpose of expelling or decomposing the nitric acid, the presence of which is injurious in the after part of the process. The addition of a little carbonate of soda will serve the same purpose. The reader will recollect that the function of the nitric acid is merely to liberate the chlorine, which is the real solvent of gold.

When the evaporation has been carried sufficiently far, three or four ounces of water should be added, after which the contents of the flask must be allowed to remain in perfect rest until the undissolved matter has completely subsided, and the supernatant fluid is quite clear. The latter is then to be carefully decanted or filtered off, and to the residue, about an ounce of fresh water must be added, left to stand till clear, again decanted or filtered; and this operation repeated five or six times, always adding the later washings to the first portion of fluid. The solution of gold now obtained will be very dilute. Add to it, therefore, a few drops of hydrochloric acid, and then introduce the protosulphate of iron, which, when the liquid is well stirred, will speedily precipitate the whole of the gold in the form of a brown powder. If oxalic acid be used for the same purpose, the liquid ought to be boiled. When the whole of the precipitate appears to have settled to the bottom, a few drops of the clear supernatant liquid should be taken out on the end of a rod, placed upon a surface of clean white porcelain, and tested with a drop of the solution of protochloride of tin. If no purple precipitate be formed, it is a proof that the whole of the gold has been thrown down. If a dark brown coloring, but still no subsidence, be produced, this will indicate the presence of platinum. If any precipitate be observed, more of the sulphate of iron or oxalic acid must be added to effect a complete precipitation of the gold contained in the solution.

When the pulverulent gold has entirely deposited, the liquid must be decanted or filtered off with the greatest precaution. Care must be taken that not the smallest particle of the gold powder is allowed to pass away with the liquid. A little hydrochloric acid, which must be quite free from any admixture of nitric acid, is then to be poured upon the precipitate. This will remove any iron or other metallic impurities without dissolving the gold. The latter is then to be washed, at least six times, with successive portions of distilled water; and lastly, it is transferred to a small porcelain or platinum crucible, in which it is heated, over a spirit lamp till the last portions of water are expelled. It ought, in fact, to be raised to a red heat, or even to be fused with a small quantity of borax and nitrate of soda, as formerly recommended, to expel the last traces of chloride of silver.

The pure gold thus obtained is then to be placed in a watch-glass or small capsule and accurately weighed in a pair of delicate scales which should be capable of turning with a difference of at least one-hundreth part of a grain. Instead of first counterpoising the dish or capsule, and then adding weights to counterpoise the gold, it is better to begin by placing the capsule *containing the gold* in one pan and counterpoising both by means of sand, or some other convenient material placed in the other. The gold is then removed from the dish and weights put in its stead sufficient to restore the equilibrium. The number of grains and fractions of a grain required for this purpose will accurately represent the weight of the gold; and in this method of *weighing by substitution*, as it is termed, any risk of error arising from the possible inequality in length of the two arms of the balance will be entirely avoided.

Supposing the gold to weigh exactly 10·75 grains and that the amount of alloy submitted to experiment was 12·5 grains, or the eighth part of one hundred, it is evident that $10{\cdot}75 \times 8 = 86$ is the percentage of pure gold present.

www.ingramcontent.com/pod-product-compliance
Lightning Source LLC
LaVergne TN
LVHW021404110826
845150LV00007B/1783

* 9 7 8 1 4 2 5 5 1 2 5 1 4 *